CHYMICO

PHY

MEDICUM,

SEU SCHEDIASMATUM

PLURIBUS ANNIS VARIIS OC
ASIONIBUS IN PUBLICUM EMISSORUM NUN
QUADANTENUS ETIAM AUCTORUM ET DEFICI
ENTIBUS PASSIM EXEMPLARIBUS IN UNUM VOLU-
MEN JAM COLLECTORUM,

FASCICULUS

PUBLICÆ LUCI REDDITUS,

Premissâ

PRÆFATIONIS LOCO

AUTHORIS EPISTOLA

ad Tit. DN. MICHAELEM ALBERT

D. & Prof. Publ. Extraordinarium

Editionem hanc adcurantem.

FAR OUT

101 STRANGE TALES FROM SCIENCE'S OUTER EDGE

disinformation®

© 2007 Mark Pilkington

Published by The Disinformation Company Ltd.
163 Third Avenue, Suite 108
New York, NY 10003
Tel.: +1.212.691.1605
Fax: +1.212.691.1606
www.disinfo.com

Design & Layout: Chika Azuma

Library of Congress Control Number: 2007935059

ISBN-13: 978-1-932857-87-0

Printed in USA

10 9 8 7 6 5 4 3 2 1

Disinformation is a registered trademark of
The Disinformation Company Ltd.

These columns originally appeared in the UK's
Guardian *newspaper in a weekly pullout science
section called Life. This collection presents the
majority of them in revised and updated versions.*

The publishers have attempted to identify the source
of all photographs and illustrations contained in
this book and to give credit where required. In many
cases, the images are in the public domain or are
considered fair use under applicable laws. However,
if we have inadvertently failed to contact you with
respect to an image that you own or control, please
contact the publishers.

Distributed in the USA and Canada by:
Consortium Book Sales and Distribution
Toll Free: +1.800.283.3572
Local: +1.651.221.9035
Fax: +1.651.221.0124 www.cbsd.com

Distributed in the United Kingdom and Eire by:
Virgin Books
Tel.: +44.(0)20.7386.3300
Fax: +44.(0)20.7386.3360
E-Mail: sales@virgin-books.co.uk

Distributed in Australia by:
Tower Books
Tel.: +61.2.9975.5566
Fax: +61.2.9975.5599
Email: info@towerbooks.com.au

**Attention colleges and universities,
corporations and other organizations:**
Quantity discounts are available on bulk purchases
of this book for educational training purposes, fund-
raising, or gift giving. Special books, booklets, or
book excerpts can also be created to fit your specific
needs. For information contact the Marketing
Department of The Disinformation Company Ltd.

FAR OUT

101 STRANGE TALES FROM
SCIENCE'S OUTER EDGE

MARK PILKINGTON

ACKNOWLEDGEMENTS

Thanks to Dr. Ben Goldacre, Mike Jay and the editors at *Life*; and to Hannah Westland and Rowan Routh at Rogers, Coleridge & White.

The photograph of the violet ray machine on page 73 is reproduced courtesy of the University of Manchester Medical School Museum. Thanks to Dr. Peter Mohr.

PART TWO:
AMAZING INVENTIONS AND SPIRITED INVENTORS

PART FOUR:
THE HUMAN CONDITION

BEFORE AND AFTER SCIENCE

Here are tales of science that was, science that might have been, science that should never have been, and science that may yet be. While this book is more of a miscellany than a comprehensive survey, it should provide a useful starting point for further exploration of science's wilder shores.

Whether we view the history of science as a grand teleological narrative or as some tangled information feedback network, the visionaries, dreamers, pioneers and charlatans described within these pages are still very much a part of that story, even if most of them have since been forgotten or ignored. History, we are taught, is written by its victors and this is as true in science as it is in warfare.

Nikola Tesla is perhaps the best-known case in point; a man who, a century ago, shaped much of the way we experience technology today, and yet is still all-too-often written out of history books. Tesla is considered problematic because of his wilder ideas, such as those about the remote transfer of energy; yet we don't write off Isaac Newton because he devoted as much of his time to the study of occultism as he did to his theories of gravitation.

When the science establishment closes ranks on the likes of Alfred Wegener, who was hounded until his death by those opposed to his theory of continental drift, or on Martin Fleischmann and Stanley Pons, who first observed desktop cold fusion, it is not methodology or reason that drives it, but fear and the pack instinct of self-preservation. Right or wrong, pioneers like these remind us that science is no monolithic, free-standing monument to progress; it is instead tightly woven into the fabric of our culture, as susceptible to fads, fashions and falsehoods as any other aspect of human endeavor.

As the natural philosopher and gentleman scientist have been replaced by the board of directors and the funding body, so science has lost much of its power to inspire. But science still needs its dreamers. The antics of our contemporaries like John Hutchison or Troy Hurtubise may cause professional scientists to snigger into their lab coats, but these are the people who have followed their dreams, no matter how absurd, and attempted to make them happen; the people whose work—caught somewhere between outsider art, magic, and the original spirit of invention—sends great ripples of amazement out into a wonder-hungry world. These mavericks' antigravity devices or invisibility rays may not work today but, somewhere down the line, stories like theirs will inspire future generations of scientists who will, in turn, transform dreams into reality.

It is this future, I hope, whose past is documented here.

Mark Pilkington

ACARUS.

PART ONE

IT SEEMED LIKE A
GOOD IDEA AT THE TIME

W *hat is there in places empty of matter?* —Isaac Newton, 1706.

Nature abhors a vacuum, and so do we. For as long as humankind has considered the universe around us, we have sought to fill its emptiness. In the fourth century B.C.E., Aristotle proposed "aether" as a name for the fifth element, or quintessence, postulated by his teacher, Plato. Aether, or ether, made up the worlds of the outer celestial spheres believed to surround our own.

By the mid-seventeenth century René Descartes was using the word to describe the medium of space. Vortices in this ether, he felt, were responsible for the accumulation of particles that formed matter and ultimately shaped all solid objects, from pebbles to planets.

A century later, the word encompassed a number of vague but related ideas about the subtle matter keeping the stars and planets in place. Many eighteenth century astronomers would invoke the aether to account for the variations and discrepancies in the motions of celestial bodies, or the way that light traveled through space.

Isaac Newton considered his version of the aether—which he described as strong, subtle and elastic while admitting that he did

not know what it actually was—to be responsible for what we now recognize as gravity and electromagnetism, as well as our own physical movements and sensations. Always mystically inclined, Newton wondered if this aether wasn't a living force—essentially, spirit—though he knew that he could never prove this to be the case.

The aether was still very much alive in the nineteenth century. Following James Clerk Maxwell's unification of electricity and magnetism, light was revealed to be just another waveband in the electromagnetic spectrum. However, Maxwell's "undulations" still needed a medium through which to travel—emptiness was not an option. The Luminiferous Aether was proposed to answer that need.

For most investigators, however, the aether vanished following the Michelson-Morley experiment of 1887, which sought, unsuccessfully, to record "aether wind" as a measurable effect of the Earth moving through the aether. To the experimenters' chagrin, their failure was taken as proof that the aether had never existed at all.

Subsequently, more sophisticated theories of subatomic physics and Einstein's space-time continuum have relegated the aether to the domains of mysticism and outré science. But its spirit lives on in dark matter, a modern manifestation of this most ancient idea.

For almost a century it was thought that flammable materials burned because they contained a colorless, odorless, tasteless and weightless substance called phlogiston (from the Greek *phlogistos,* flammable)—the matter of fire.

The theory was developed in late seventeenth century Germany by two university professors, Johann Joachim Becher and then Georg Ernst Stahl, who outlined his version in *The True Theory of Medicine* (1708). Expanding on the ideas of the sixteenth century Swiss chemist and physician Paracelsus, Becher suggested that metals and minerals were compounds which, when burned, released *terra pinguis* (meaning fatty, flammable earth) into the air. This left behind the metal's true form, the *calx,* composed of *terra lapida* ("stony earth"—giving a metal body), and *terra mercuriallis* ("mercurial earth"—giving it weight and color).

Stahl extended Becher's theory, replacing *terra pinguis* with phlogiston in 1700. Phlogiston was released into the air by the combustion of flammable matter and the respiration of living organisms. These processes formed phlogisticated air, which was then absorbed by plants. Charcoal, which left little residue when burned, was considered to be almost pure phlogiston, and when burned with a metal's *calx,* would restore it to its familiar compound state. Rusting iron, according to the theory, was slowly releasing its phlogiston into the air and so returning to its original, elemental state.

Among phlogiston's supporters was the scientist and theologian Joseph Priestley, inventor of the carbonated drink and discoverer of numerous key gases. One of these he described as being "an air five or six times as good as common air"; indeed, a candle placed inside it would burn extra brightly and a mouse breathing it lived for twice as long as it did in ordinary air (*Experiments and Observations on Different Kinds of Air* Vol. II, 1775). Because things burned so well in this new type of air, Priestly called it "dephlogisticated air" and informed the French chemist Antoine Lavoisier.

Lavoisier took Priestley's dephlogisticated air into the laboratory, where he reached conclusions of his own. These he presented to the French Academy of Science on September 5, 1775—the day phlogiston died. "The dephlogisticated air of Mr. Priestley, is... the true combustible body," said Lavoisier, "...there is no longer need, in explaining the phenomena of combustion, of supposing that there exists an immense quantity of fixed fire in all bodies which we call combustible."

Lavoisier would rename Priestley's dephlogisticated air "pure air" and later, "oxygen."

aving trained as a physician in Vienna and practiced as a doctor in the Saxe German mining town of Hettstedt, Samuel Hahnemann became increasingly disillusioned with the medical establishment's attitude towards the treatment of patients, which he considered to be vague and dangerous.

In 1790 he translated the distinguished Scottish physician William Cullen's *Materia Medica* into German, but took issue with Cullen's explanation of how Peruvian Cinchona bark (in which quinine would be isolated 28 years later) cured fever. So he decided to experiment with it on himself.

After several trials, in which he induced fever and palpitations, Hahnemann established the theory that the outward symptoms of a sickness were a reflection of the "untuning" of the *vis vitalis*—the life force thought to differentiate organic from inorganic matter—and that the effects of a substance on the healthy organism revealed what symptoms it could cure in the unhealthy organism. Working from the principle *Similia Similibus Curentur,* "like cures like," Hahnemann began systematically to experiment on himself and members of his family, noting the effects of numerous different substances on their healthy bodies.

In 1805 he published a directory of the symptoms induced by different medicines—the first of its kind—and developed his ideas into a practical system of treatment, which he called homeopathy. Although Hahnemann himself was subjected to a series of professional attacks by medical practitioners, homeopathy grew in popularity with Europe's patients. His method's moment of triumph came in 1813, when Hahnemann successfully treated a number of patients in Leipzig during a nationwide outbreak of typhus fever.

The central tenets of homeopathy—that like cures like and the more a solution is diluted the more potent it becomes—have changed little since Hahnemann's time. The practice achieved its greatest notoriety in 1988, when *Nature* published a paper by the French immunologist Jacques Benveniste. In it he observed that white blood cells still demonstrated an allergic response to water in which allergens had been diluted down to minute homeopathic doses. Benenviste's conclusion, that water somehow retained a "memory" of the molecules it once contained, caused a storm of controversy. A subsequent investigation by a *Nature*-appointed team attempted to discredit his findings, criticizing the working environment of his lab and his statistical analysis; but the team found no evidence of fraud.

The subject remains deeply controversial. Researchers on opposing sides of the argument regularly claim to have successfully replicated Benenviste's findings, or to have disproved homeopathy entirely. A 2005 survey published in the leading medical journal *The Lancet* looked at 100 homeopathic trials and found the results to be no better than those for placebos. It has been suggested that measured improvements in homeopathic patients arise largely from the experience of therapy itself—the time and attention spent on the patient by the homeopathist—rather than the mysterious little white pills prescribed as medicine.

But an improvement is still an improvement, however it works, and despite repeated attempts to undermine his theories, today Hahnemann's 200-year-old medicinal art accounts for 0.5 percent of the world's pharmaceuticals market. ⌗

004 FOOL'S GOLD: ALCHEMY

May 25, 1782. In the culmination of a month's worth of startling experiments, 15 prestigious observers, among them several Lords and Fellows of the Royal Society, watched keenly as the 24-year-old chemist James Price mixed mercury with a tiny amount of a mysterious red powder. The substances were heated in a crucible for several minutes then allowed to cool. In their place appeared a yellowish metal, later identified by an Oxford goldsmith as constituting a high standard of gold.

Generally speaking there are two schools in the study of Western European alchemy. The more literal approach sees the sixteenth and seventeenth century alchemists as early chemists, predecessors of the sort of people James Price had recently joined as a Fellow of the Royal Society. The other school sees the alchemists in a more mystical light, and their descriptions of physical experiments as an allegory for a process of spiritual transformation.

Such a distinction would have meant little to many of the early alchemists, for whom spirit and matter were intrinsically connected, but by the time of Price's experiment, the practice had only one aim—turning base metals into gold.

News of Price's transformation spread quickly and caused a sensation; a sample of his gold even received the personal approval of King George III. Oxford University awarded him a doctorate and he wrote accounts of his experiments for the newspapers.

Within the Royal Society, however, all was not well: many important members, not least the president, Joseph Banks, were highly skeptical of his claims, but their demands that Price reveal his methods, and the composition of his miraculous red powder, fell on deaf ears. As the pressure upon him grew, he finally relented and agreed to a conclusive demonstration.

In early August 1783, with his reputation now close to tatters, Price invited three members of the Royal Society into his home. Accounts of what happened next vary, but at some point Price downed a glass of highly poisonous laurel water that he had distilled himself, and within hours, or maybe even minutes, he was dead.

Was Price simply a fraud who feared exposure? Was he suffering delirium as the result of mercury poisoning? Or, perhaps, as alchemist and historian Guy Ogilvy has suggested, the mystery powder was passed to him by another alchemical adept, meaning that Price was unable to reveal its secrets because he didn't actually know them himself.

The answer, and perhaps the secrets of alchemy, died with James Price, just as a new age of science was born. ☿

A diligent gentleman scientist with a passion for electricity, Andrew Crosse was known as the "thunder and lightning" man around his home in the Quantock Hills, Somerset, where he also held a seat as an English Member of Parliament.

In 1836, Crosse was attempting to create artificial crystals by dripping a chemical solution through an electrified stone from Mount Vesuvius. After two weeks, Crosse noticed tiny white spots appearing on the surface of the stone, which was submerged in a mixture of water and hydrochloric acid. On the eighteenth day, filaments began to emerge from the spots and by the twenty-sixth day the spots had taken on a startling form. "Each figure," wrote Crosse, "assumed the form of a perfect insect, standing erect on a few bristles which formed its tail." On the twenty-eighth day, to Crosse's astonishment, the insects began to twitch their legs. Within weeks, a hundred had formed and, once rescued from the water, scuttled across his workbench in search of shelter.

The tiny "porcupine"-like creatures were identified as mites of the genus *Acarus*—it was initially thought they might represent a new species—perhaps *Acarus crossii* or *Acarus galvanicus*. Wishing to experiment further before making a public announcement, Crosse tried to keep his discovery quiet. But word soon got out and Crosse found himself labeled "a reviler of our holy religion," "a disturber of the peace of families," and denounced as a Frankenstein-like madman.

He had some surprising defenders, however. In 1837 the distinguished electrical pioneer Michael Faraday told the Royal Institution of Great Britain that he too had observed the appearance of mites during some of his experiments, though he didn't think that they were actually born of electricity. Nor, for that matter, did Crosse, though this did nothing to quell the vitriol of his detractors. "I have never ventured an opinion as to the cause of their birth, and for a very simple reason—I was unable to form one. The most simple solution of the problem which occurred to me, was that they arose from ova deposited by insects floating in the atmosphere," he wrote in 1837.

Crosse would have been working in homely conditions, and it seems most likely that his mites were of the common dust or cheese varieties, whose eggs had already contaminated his equipment before the experiments began. At least one fellow investigator, William Henry Weekes of Sandwich, England, repeated and expanded upon Crosse's work, noting that the number of mites was proportionate to the amount of carbon in the mixture, but his findings attracted little attention and Crosse remains to this day something of an electrical heretic.

As a fabulously wealthy industrialist running one of Europe's most successful steel firms, Baron Karl Freiherr von Reichenbach could afford to indulge his insatiable scientific curiosity. Amongst his discoveries were paraffin wax (1830) and creosote (1832), while his blast furnace design revolutionized the steel industry. The baron was also a keen collector of aeroliths, what we today call meteorites, defying the wisdom of his time to propose that the rocks had fallen from the sky.

But what really fascinated Reichenbach was the phenomenon of sleepwalking and other night terrors. After a meticulous investigation, during which he interviewed hundreds of sufferers, the Baron drew the conclusion that some external force was driving these nocturnal dérives. The "sensitives," as he called them, represented all ages and all walks of life, but what were they sensitive to? Nothing if not a painstaking researcher, the doctor considered all the avenues known to science—magnetism, electricity, chemistry, mesmerism—before deciding that it must be something else, a hitherto unknown "occult force," which he termed Od (from Odin—being transcendent and all-pervading like the Norse god, and *odos*—path in ancient Greek).

Initially Reichenbach considered light to be the main carrier of Od and ran hundreds of fantastical experiments in his castle, Cobenzl, outside Vienna. Sensitives in specially darkened rooms described seeing flickering flames, human faces and rainbows of color as the Od flowed through wires attached to metal receptor plates fastened to the castle's outer walls. There were physical effects too; sometimes Od was benevolent and nourishing, at others a harmful irritant. Here, felt the Baron was the answer to the sleepwalking mystery: nourishing Od rays were drawing the somnambulists out of their homes to bask in the healing light of the Moon.

But it didn't stop there. Reichenbach and his sensitives found that everything—light, minerals, metals, organic matter, human bodies, even sounds—gave off Od. "We live in a world full of shining matter," he proclaimed. Od was everywhere.

You won't be too surprised to learn that the scientific and medical establishments either ignored or ridiculed Reichenbach's experience-led research. But he had many dedicated supporters and continued to explore his luminous new world until his death, at the age of 81, in 1869. 〇

"The time of deliverance has come at last," cried *New Era*, an American spiritualist newssheet, in July 1854. It hailed the "birth" of the New Motive Power, a miraculous mechanical device described by its 50-year-old "father" John Murray Spear as "Heaven's Last Gift to Man" and the "New Messiah."

Spear, an ardent spiritualist and fervent abolitionist, was in psychic contact with the "Band of Electricizers," a committee of scientifically minded spirits headed by Benjamin Franklin. As well as plans for an improved sewing machine, an electric boat and an intercontinental telepathic network, the group showed him how to boost signals to the spirit world by covering himself in copper and zinc batteries. Their tip worked and he was soon downloading blueprints for the New Messiah.

A child-sized anthropomorphic marvel, the New Motive Engine had steel arms and spheres, zinc and copper plates for its brain, antennae to draw power from the skies, a chest of wires, bars, chemicals and magnets and earthing rods for legs. Its construction, at High Rock, outside Lynn, Massachusetts (a site chosen by the Electricizers) took nine months, during

which time Spear was assisted by *New Era*'s editor and "the Mary of the New Dispensation," generally thought to be Mrs. Spear.

As the machine neared completion Spear, wearing a suit of gemstones and metal strips, entered a deep trance and created a psychic "umbilical" link to his creation. The New Mary, meanwhile, declared herself pregnant and, on June 29, 1854, lay next to the machine for two hours, overcome by labor pains. When these ended she laid hands on the robot-child, which, according to those present, began to move listlessly, like a feeble newborn infant. There was some debate amongst the witnesses as to what this movement consisted of, but it was soon decided that a change of air would boost its spirits and inject some vigor into its wiry veins. Sadly, after being relocated several hundred miles away to Randolph, New York, the machine, according to local legend, was destroyed by an angry, superstitious mob. Unfortunately there are no surviving records of its violent demise.

Spear himself moved on, his mechanical child seemingly forgotten, and progenated a human child with a new wife, the women's rights activist Caroline Hinckley. At the spirits' behest he developed a controversial doctrine of female emancipation, communal living and free love, which he preached until, so instructed by the spirits, he retired in 1872. ☿

"**B**efore my eyes a large spherical mass, about eight inches in diameter, emerged from the vagina and quickly placed itself on her left thigh while she crossed her legs. I distinctly recognized in the mass a still unfinished face, whose eyes looked at me."

Baron Albert von Schrenck Notzing, the respected Munich psychiatrist and physician in whose book, *The Phenomena of Materialisation* (1923), this passage appears, became fascinated with mediumistic phenomena while conducting early hypnotic experiments in the late nineteenth century. The Baron studied the medium Marthe Beraud, known as Eva C, for over a decade, though he didn't actually witness her more spectacular manifestations, instead taking—and apparently believing—testimony from her French adoptive mother Mme. Bisson.

What the Baron calls "mediumistic teleplastics" would become better known as ectoplasm, (from the Greek, "exteriorized substance" or "formed outside of the body"), a mainstay of physical mediumship demonstrations of the later nineteenth and early twentieth centuries. Emerging from every orifice on the medium's body, ectoplasm would first manifest in the shape of drops or a thin thread, before expanding to take on fuller shapes, whether human, animal or entirely abstract. Sometimes viscous like albumen, sometimes more solid and rubbery and at other times netted like muslin, ectoplasm was said to be extremely sensitive to touch and to sunlight; hence conveniently enough, the preference of mediums to perform undisturbed and in near total darkness.

Those mediums, mostly female, who did subject themselves to scientific investigation underwent an often undignified ordeal of restraint and invasive examination. Some, like Beraud, managed to convince researchers that their psychic expectorations were indeed paranormal in origin; many more were exposed as frauds.

In 1931, the Scottish medium Helen Duncan—in 1944 the last person to be prosecuted under the UK Witchcraft Act of 1735 after she revealed naval secrets during a séance—was examined in the laboratories of famed psychical researcher Harry Price. He was unimpressed by what he saw. Duncan's ectoplasmic emissions consisted of cheesecloth, safety pins, rubber gloves and an organic substance that may have been egg white. Price wrote up his findings in *Regurgitation and the Duncan Mediumship* (1931), whose title reveals the uncomfortable method by which Duncan produced her sticky spirit forms.

The psycho-spiritual effluvia of a more innocent age, ectoplasm was unable to survive such harsh exposures and, like crinolines and magic lantern shows, can be considered amongst the lost arts of the Victorian era. ⌁

ené Prosper Blondlot was a distinguished physics professor at the University of Nancy, France, when, in 1903, he announced a discovery that he hoped would make him famous. Instead, it brought his career to an abrupt end.

While attempting to polarize the newly discovered X rays, Blondlot stumbled across what he believed to be another form of invisible radiation. He named them N rays, after their hometown. The N rays displayed a number of curious properties: when refracted through an aluminum prism to cut out visible light, the rays caused a thread coated with fluorescent calcium sulphide to glow; an electric spark also seemed to glow brighter in their presence; heat increased their effect, while loud noises dissipated it. Unlike other forms of radiation, such as natural light or X rays, which become diffuse when shone through a pinhole camera, N rays retained a high level of resolution. It seemed that everything Blondlot examined gave off N Rays, including human bodies, the only exceptions being green wood and certain treated metals.

Blondlot excitedly announced his discovery to the French Academy of Sciences, and published a paper in *Nature*. Fellow scientists responded with enthusiasm at first, some even claiming the discovery as their own. But when several famous figures, including Lord Kelvin and William Crookes, attempted to replicate the N ray experiments, none could reproduce Blondlot's findings.

Smelling a rat, *Nature* dispatched Robert W. Wood, a leading optical physicist, to investigate. Blondlot showed him how an electrical spark glowed more brightly when exposed to N rays; Wood, however, saw nothing. Shown photographs of sparks getting brighter in the rays' presence, Wood remained unconvinced. Finally, as Blondlot set up the apparatus for another experiment, Wood surreptitiously removed the crucial aluminum prism, without which—according to Blondlot—the rays could not be seen. As the experiment continued, Blondlot and his assistant insisted that the N ray spectrum was still visible. The game was up.

Wood was convinced that the N rays were a figment of Blondlot's imagination, and wrote as much when he described his experience in *Nature* (September 1904). Blondlot emerged from the investigation as either a fraud or a fool, and his career never recovered from such public ignominy.

While a tiny few still hold out for their existence, for most commentators N rays serve as a reminder of science's susceptibility to human fallibility and, more positively, of its ability to correct its own mistakes. And as for Blondlot...well at least he hasn't been forgotten.

I n 1882, Dr. Albert Abrams returned to his native California having received his medical doctorate from the prestigious University of Heidelberg in Germany. There Abrams had gained a reputation for diagnosing illnesses by a system of physical taps on the body, and within a decade of being back in America, he was made professor of pathology at Cooper Medical College.

Extending his diagnostic methods, Abrams discovered that different diseases caused tissues to vibrate at different frequencies and developed a device, the Reflexophone, to measure them. He next theorized that by turning these frequencies around and blasting them back at the affected tissue, one could effectively cleanse the tissue of disease. This he did with his next invention, the Oscilloclast, through the use of which diagnoses and even "remote" treatments could be made from a patient's blood or hair.

Abrams promoted his new system, Electronic Reactions of Abrams (ERA), with a 1916 book, *New Concepts in Diagnosis and Treatment*, and by 1923 over 3000 doctors were using Oscilloclasts at work. The income from his invention made Abrams a multi-millionaire and garnered him many influential supporters, but by the time of his death in 1924, attacks on his ideas had grown increasingly vociferous. *The Lancet* claimed that samples of healthy chicken blood sent to ERA practitioners had been diagnosed as having major diseases, while a *Scientific American* report reached the damning conclusion that: "At best, [ERA] is all an illusion. At worst, it is a colossal fraud."

ERA disappeared almost entirely from the medical arena, but the system continued to evolve, most notably with the refinements made by Dr. Ruth Drown, who renamed it Radionics. A brief reprise in the early 1950s saw Radionics devices being employed in agriculture. A company called UKACO (made from the founders' names, Upton, Knuth, Armstrong Company) claimed that Abrams' technology could kill pests in farmers' fields, simply by sprinkling a tiny amount of pesticide onto a photograph of the afflicted area and placing it in a modified Oscilloclast box. Although perhaps closer to sympathetic magic than science, several farms testified to the system's effectiveness. The US Department of Agriculture disagreed however, and closed UKACO down.

While you won't find Radionics devices being used in your local hospital today, Abrams' technologies retain a committed following. Claims of miraculous healings abound, alongside sinister rumors of Radionic murders. Call it magic, madness or mysticism, say the children of the new ERA, just don't say it doesn't work. ⌐

I magine if Creationists took over the education system. "It'll never happen," you say. But something very similar already did, in Stalin's Soviet Union.

Born into a Ukrainian peasant family in 1898, Trofim Denisovich Lysenko first made the news in 1927, when *Pravda* ran a story on the agronomist's radical new crop management ideas. Lysenko employed a technique by which winter crop seeds, after being chilled and soaked, could be planted and grown in summer, thus potentially doubling the annual harvest. Although the discovery was not his own, Lysenko named it "yarovizatsiya," vernalization, and was made head of a vernalization department at the Institute of Genetics and Plant Breeding.

As the vernalization movement grew, so did Lysenko's ideas. Based on some extremely dubious experiments, he developed a theory around the neo-Lamarckian ideas of fruit tree hybridizer Ivan Michurin. Lamarck—who first coined the term "biology," and whose evolutionary theory preceded Darwin's—proposed that a species' inherited traits developed as a response to its environment. A classic example was the giraffe, whose long neck grew after its ancestors began stretching to eat high leaves off trees.

Lysenko argued that this interpretation of evolutionary processes best corresponded to Marxist theory—it was, after all, concerned with struggle and radical change. The Stalinist regime agreed. Eventually, aided by the Communist philosopher and ideologue Issak I. Prezent, Lysenko rejected the existence of genes entirely. Such was his status that those scientists who disagreed with his unorthodox ideas renounced the error of their ways or faced the consequences. Several prominent geneticists disappeared into gulags.

By 1948, Lysenkoism represented the Soviet Party line on evolution. With Stalin's support he became a Hero of the Soviet Union and a vice president of the Supreme Soviet. Although a subsequent, massive reforestation campaign based on Lysenko's ideas was a huge disaster, it was not until Stalin's death in 1953 that his star began to fade.

Personally criticised by Khrushchev, it soon began to emerge that Lysenko had faked some of his experiments to support his theories. By 1965 he had lost his directorship at the Institute of Genetics and was thoroughly discredited, but he remained an agricultural adviser to Khrushchev and died in 1976.

The Lysenko story serves as a powerful warning against the collusion of ideology and science, a threat that remains very much with us today.

Austrian biologist Paul Kammerer's paper *The Inheritance of Acquired Characteristics* caused uproar when it was published in 1924, with its author celebrated as "the new Darwin."

Kammerer claimed to have experimentally demonstrated the evolutionary theory of the nineteenth century naturalist Jean-Baptiste Lamarck who, in 1809, had stated that repeated use of an organ by a species caused that organ to evolve and develop in its descendents. Lamarck's theory was disproved by the German August Weissman in the 1880s, when he found that, no matter how many mouse tails he chopped off, their progeny continued to sport them.

An expert on a number of amphibious and marine species, Kammerer chose the midwife toad for his experiments. These toads mate on land rather than in water, and lack "nuptial pads," the small dark bumps on the male forelimbs of most toad species, which provide them with extra grip while mating underwater. The male midwife toad carries the female's eggs on his back until they hatch in water, hence the name.

The experiment involved separating the toads into two groups; one was forced to live permanently on dry land, the other in a tank

containing small islands, where the air temperature was uncomfortably high. After a few generations, Kammerer claimed that the landlubbers stopped carrying eggs on their backs (because they no longer hatched in water), while the water babies developed nuptial pads to facilitate better water sporting.

Kammerer's findings were hailed as "the greatest biological discovery of the century," but his triumph would be short lived. In *Nature*, August 7, 1926, Dr. G. K. Noble of the American Museum of Natural History made a devastating announcement. Studying the last remaining Kammerer toad specimen, passed to him by a colleague, Noble found that its nuptial pads were strangely smooth, lacking telltale epidermal spines. What was more, the black coloring seemed to be under the skin, not in it, and the toad's wrist was slashed. When he opened the wound, black ink trickled out—this was a tattooed toad.

Amid cries of hoax, Kammerer protested his innocence, declaring sabotage. But the shame, coupled with the pain of a failed romance, was too much for him.

"I find it impossible to survive my life work's destruction," he wrote in a letter published in *Science*. "I hope to find tomorrow sufficient courage and fortitude to end my wretched life." The next day his career, and his life, was over. ☐

erhaps the twentieth century's most infamous scientific heretic, Wilhelm Reich's career saw him go from being a favored protégé of Sigmund Freud to dying a sad and ignominious death in a US federal jail.

After serving in the Austrian army during World War I, Reich studied medicine and began giving seminars on sex and psychology in Vienna, drawing the attention of Freud. The two developed a close relationship, working together for almost ten years. During this time Reich extended Freud's ideas about the libido to suggest, in his book The *Function of the Orgasm* (1927) that the lack of orgastic energy—i.e. a lack of orgasms—lay at the root of all neurosis. His extreme ideas lead to a split with Freud and Reich grew increasingly fascinated with Communism, before turning angrily on the "groupthink" of the Communists and Nazis in his book *The Mass Psychology of Fascism* (1933). Reich then extended his sexual theories into the realm of politics, culminating in *The Sexual Revolution* (1936) in which he espoused sexual freedom and pleasure as the keys to avoiding the dysfunctional societies he saw developing around him.

Moving to Norway, Reich returned to medicine and claimed the discovery of "bions," blue organisms produced by rotting matter. These bions occupied the space between living and inanimate life, and also seemed to destroy germs and bacteria. Reich also noticed that where his skin came into close contact with bions it would tan, leading him to surmise that they gave off a kind of radiation, which he would term "orgone." Orgone was the force of life itself, the sexual energy given off during orgasm, and Reich spent some time carefully studying the energetic emissions of people at different stages of sexual arousal.

Considered too outrageous for Norway, Reich took his research to America, setting up shop in Rangely, Maine. Working in his new labs, which he named Orgonon, Reich came to believe that orgone was everywhere, and that its natural flow was the very pulse of life on Earth. Blocking its flow in the individual could lead to tension, neurosis and physical deterioration, so Reich set about designing a tool for generating and amplifying the precious substance. The orgone energy accumulator—ORAC—was a wardrobe-sized box, composed of sheet metal on the inside and wood or other organic materials on the outside, that would bathe anything within it in beneficial orgone energies. Reich experimented with cancer patients and claimed that, with regular orgone treatments, their tumors reduced in size; they were relieved of pain and survived considerably longer than initially expected.

But just as there was orgone, there was anti-orgone or DOR (Deadly Orgone Radiation), largely created, Reich felt, by atomic testing. He developed another device, the cloudbuster, which extracted DOR from the air and was also capable of modifying weather patterns and creating rain and snow on occasion. He also became obsessed with UFOs, which he concluded were spreading DOR to facilitate the end of humankind.

After some negative press, rumors spread that Reich was heading a sex cult, and the US Food & Drug Administration took an interest in his work. The FDA's main accusation, that Reich claimed his ORACs could cure cancer, was unfounded, and when the FDA summoned him to court, the already paranoid Reich ignored their order, smelling a conspiracy. The court ordered him to stop selling the orgone accumulators, but Reich persisted, seeing it as his duty. A second summons resulted in a two year prison sentence and, worse, the destruction of all his ORACs, his laboratory equipment, papers and books, of which over six tons were burnt. A broken man, Wilhelm Reich died in prison on November 3, 1957, two months before he was to be paroled.

In death Reich has become a fringe science martyr, inspiring conspiracy theories, books, music and a wild Yugoslavian documentary film *WR: Mysteries of the Organism* (1971). Seminars are still held at his Orgonon laboratory, which also functions as a museum. ⛘

he notion that large, hitherto unidentified creatures may exist in our oceans and wildernesses is one that most people are comfortable with. But could colossal, primitive lifeforms, invisible to human eyes, also populate our skies?

Trevor James Constable—sailor, aircraft historian and scientific iconoclast—certainly thinks so, and he's not alone. Inspired by Wilhelm Reich's orgone energy, Ruth Drown's radionics, the writing of Charles Fort and Arthur Conan Doyle's story *The Horror of the Heights*, Constable became convinced that the UFOs he heard so much about in the 1950s weren't alien spacecraft, but living beings.

Armed with a camera fitted with high-speed infrared film and an ultraviolet filter, Constable set out to reveal these sky beings to the world. His photographs certainly show something. To the untrained eye they look like stains or discolorations produced during the developing process. Stare long enough, however, and they take on the appearance of floating, zeppelin-sized amoebas.

In his 1975 book *The Cosmic Pulse of life*, Constable calls them "critters." "As living organisms," he writes, "critters appear to be an elemental branch of evolution probably older than most life on earth, dating from the time when the planet was more gaseous and plasmatic than solid… Like fish, I estimate them to be of low intelligence. They will probably one day be better classified as belonging to the general field of macrobiology or even macrobacteria inhabiting the aerial ocean we call the sky."

The critters are, thankfully, usually invisible to us, existing for the most part in the infrared range of the electromagnetic spectrum. When they do occasionally stray into our frequency band, they are mistakenly identified as flying machines.

Constable's theory—an engaging synthesis of science, ufology, occultism and cryptozoology—struck a chord with many readers at the time, convincing one zoologist to name the creatures *Amoebae constablea*, after their discoverer.

Thirty years on, even ufologists consider Constable something of a fringe character. However, his spirit lives on in lesser phenomena such as "rods"—alleged airborne lifeforms that can only be captured on digital camcorders—and "orbs," balls of light, beloved of ghost hunters, found predominantly in digital images. These modern variations have been effortlessly trapped, analyzed and ultimately dismissed as dustmotes, fast-moving insects and in-camera artifacts by digital debunkers. Meanwhile, somewhere up there, Constable's skywhales continue to roam free.

Papal portrait artist and one-time opera singer Friedrich Jürgenson was recording birdsong at his country home outside Stockholm in June 1959, when he picked up something unusual on his tape recorder. Listening to the tape he heard the birds fading away to be replaced by a voice saying, "Friedel, can you hear me? It's mammy." It was his mother's voice; Friedel was her nickname for him. But she had been dead for several years.

Earlier that year Jürgenson had received taped messages from "Central Investigation Station in Space," alleged space beings who claimed to be conducting "profound observations of mankind." Previously the same June, again when taping birdsong, he had picked up a trumpet solo and a man's voice speaking in Norwegian. Jürgenson first considered the voices to be those of extraterrestrials—this was the Sputnik era, after all—but his mother's message finally convinced him that he was recording the voices of the dead.

At first using a microphone and tape recorder and later, at the suggestion of the voices, using a radio, Jürgenson would ask questions out loud and record the dead air afterwards. Playing back the tapes, often at slow speeds, he would hear voices, usually speaking in a mixture of languages—Jürgenson himself spoke ten—which he labeled "polyglot," meaning "many tongues."

The idea of radio communication with the dead was nothing new. Bell, Edison and Marconi all believed that their respective inventions would eventually make it possible. But Jürgenson claimed actually to have done it, and his recordings drew the attention of the world's press and, of course, parapsychologists. His most influential protégé was the Latvian psychologist Konstantin Raudive, whose book and record *Unhoerbares wird hoerbar (The Inaudible Made Audible),* translated into English in 1971 as *Breakthrough,* became hugely popular. A new wave of researchers labeled the voices Electronic Voice Phenomena (EVP), the blanket term used for the phenomenon today.

Many EVP recordings clearly comprise stray radio signals recorded between broadcast stations. Following the work of psychologist Diana Deutsch, artist and researcher Joe Banks has suggested that the EVP "voices" result from our tendency to form meaningful patterns from random sensory data. Regardless, EVP is still a popular method for afterlife communications, and is currently used by some bereavement coun-

selors. Meanwhile devices like American George Meek's Spiricom allegedly facilitate direct, instantaneous two-way communications with the other side.

Can Electronic Text Phenomena be far behind?

016 VELIKOVSKY: THE DAY THE EARTH STOOD STILL

A polymath, professional psychoanalyst and life-long friend of Albert Einstein, Immanuel Velikovsky created an international furor with his 1950 book *Worlds in Collision*.

Comparing accounts in the Bible's books of Exodus and Joshua and a contemporaneous Egyptian document, *The Admonitions of Ipuwer*, Velikovsky concluded that they chronicled a series of catastrophic events that occurred during the second millennium B.C.E. These include the parting of the Red Sea, the pillar of smoke and fire, the eruption of Mt. Sinai, the various plagues described in Exodus and, recorded in the Book of Joshua, a huge meteorite fall and a period of time when the Sun stood still in the sky. Velikovsky theorized that these events signaled the emergence of the planet Venus as it broke loose from Jupiter and swept through our solar system in the shape of a comet. This Venus-comet, he suggested, almost collided with Mars and the Earth before settling into its current orbit, causing our own planet to stop rotating for a brief period.

Worlds in Collision rode the bestseller lists for twenty weeks, but the response from the astronomical establishment was devastating. Perhaps because of his lack of academic astronomical credentials, Velikovsky was vilified and denounced as a charlatan and a fool. One conspiracy theory even suggested that Velikovsky, a Russian Jew, was being fed disinformation by Russian scientists seeking to destabilize Western academia. Some of his supporters in both science and publishing lost their jobs, while academics continued to organize boycotts of his lectures and books right up until his death in 1979.

Although much of his evidence has since been proven to be flawed, Velikovsky's basic premise, that millennia ago some kind of catastrophe took place in our solar system, has gained some acceptability. By studying the motion and makeup of asteroid belts, astronomer Tom van Flandern has posited that there may have been two extra planets in our solar system at one time, and that Mars was once a moon that exploded from a now-vanished planet.

Velikovsky is unlikely to join Bruno and Galileo in the pantheon of redeemed cosmological heretics, but his admirers continue to defend his name, which is assured a prominent footnote in the annals of science.

The discovery of more possible planets in our solar system always raises pulses in the cosmological community, but for a few anxious sky watchers, the news may herald our own planet's impending annihilation.

The story begins in the mid-nineteenth century, when French and English astronomers studying discrepancies in the orbits of Saturn, Jupiter and Uranus predicted the existence of another planet, later named Neptune. But the orbital eccentricities remained, leading Percival Lowell to propose the existence of "Planet X." It was while hunting this elusive planet that Clyde Tombaugh found Pluto, whose planetary status was recently downgraded to "dwarf planet" (which

sounds more like a '50s science fiction film, and certainly one that this author would watch) along with the larger Eris, discovered in 2003. Fresh data gathered by the Pioneer and Voyager probes has ruled out mainstream astronomical speculation about

Planet X. But that doesn't mean it has gone away.

Through six books, beginning in 1976 with *The 12th Planet*, Russian-born researcher Zechariah Sitchin has evolved a vastly complex mythology that spans half a million years and rewrites human origins. Based largely on his (entirely incorrect, according to at least one scholar) interpretation of a single Sumerian seal, VA 243, Sitchin proposes that Earth was colonized 450,000 years ago by an alien race fleeing their dying home planet, Nibiru. Known to different ancient cultures as the Annunakai, the Nephilim and the Elohim, about 300,000 years ago the aliens genetically upgraded monkeys to create a perfect slave race—humans.

Nibiru, however, is still out there, orbiting the Sun every 3600 years. This orbit brings it perilously close to Earth, resulting in cataclysms like the great deluge documented by so many mythologies.

Sitchin's grandiose mix of myth, fantasy and history has earned him a modest global following and, while he himself has never heralded Nibiru's imminent return, others have. For several years leading up to May 16, 2003, Californian Nancy Lieder produced reams of digital dementia warning that Nibiru was coming, and with it global devastation. In one of her last media appearances she provoked outrage by saying she had killed her two young German Shepherds to prevent them suffering during The End and, more pragmatically, so that she could eat them afterwards.

May 16, 2003 saw a beautiful lunar eclipse, but no apocalypse. Lieder herself has since disappeared, but Planet X, say believers, is still out there. ⬚

018 HARMONIC 33: THE INTERNATIONAL GRID

I n 1952, New Zealand airline pilot Bruce Cathie's life changed forever. From Auckland City's Manukau Harbor he watched amazed as a bright white light, accompanied by a smaller red light, "carried out maneuvers that no known man-made vehicle could accomplish at that time." Cathie had seen his first UFO.

Collating all the information he could on the subject, Cathie discovered the work of French ufologist Aimé Michel who, in the early 1950s, had proposed that UFOs traveled the world by following straight lines between specific waypoints. As a pilot, the theory made perfect sense to Cathie, who began to plot UFO "flight paths" around New Zealand. Before long, he had charted a complex, radial grid system over the entire country. But the UFOs were being spotted all over the world: it was time to expand.

The key to the mystery emerged on August 29, 1964, 12,808 feet (3904 meters) beneath the ocean West of Cape Horn, when the underwater camera of the Oceanographic Research Vessel Eltanin photographed a strange object. Described by a New Zealand newspaper as looking "something like a complex radio aerial," the Eltanin object was just what Cathie was looking for—an alien navigation beacon. Using a plastic sphere he was able to align his NZ energy grid and the "aerial" precisely. He now had his global grid. But it didn't stop there.

"Finally," wrote Cathie in 1994, "after years of work, I discovered that I could formulate a series of harmonic unified equations which indicated that the whole of physical reality was in fact manifested by a complex pattern of interlocking wave-forms. I gradually found that the harmonic values could be applied to all branches of scientific research and atomic theory."

Over a series of books, beginning with *Harmonic 33* (1968), Cathie rewrote the rules of space-time; worked out that atomic bombs could only be detonated at a "specific time in relation to the geometrics of the solar system" and even improved Einstein's theory of relativity to create the Harmonic Unified Field Equation, $E = [c + \sqrt{(1/c)}]\, c^2$ which, simply put, demonstrates that "all of creation is light."

Although his harmonic theories had left the Earth behind long ago, Cathie had, however overlooked one crucial piece of information: in 1971 the Eltanin object was identified as a fine, upstanding specimen of *Cladorhiza concrescens*—a sea sponge.

Did our quadrupedal, pre-hominid ancestors lead a semi-aquatic existence before emerging as the hairless bipeds that eventually became us?

So suggests the Aquatic Ape Hypothesis (AAH), an alternative human evolutionary model that has proved enduringly popular over the past three decades.

Alister Hardy, a distinguished marine biologist, first presented *Homo aquaticus* to the British Sub Aqua Club in 1960, following it up with several *New Scientist* articles. He'd originally conceived of the AAH in 1927, after watching sea mammals being dismembered as a young man, but decided to keep it to himself, not wishing to jeopardize his career prospects.

The basic premise is that our tree-dwelling ancestors began to hunt on seashores, eventually entering the water to catch crabs and fish. Several evolutionary developments resulted from this transition: the apes lost the bulk of their body hair, those strands that remained aiding the flow of water over their bodies as they swam; they developed a more upright, bipedal posture, supported by the water; their fingertips grew more sensitive as they felt around for food; they grew a layer of subcutaneous fat—found only in other aquatic mammals—to keep themselves insulated; they become adept swimmers, and they used stones as tools to crack open shells. Their remains were, rather conveniently, swept out to sea, explaining the lack of sea monkey fossils.

While it raised a few eyebrows, Hardy's theory had been largely forgotten when it re-emerged with a vengeance in 1972, thanks to Elaine Morgan, a successful TV scriptwriter with no scientific background. She adopted the AAH as the central premise of her feminist anthropological treatise, *The Descent of Woman*, adding a few touches of her own, most importantly that, to give their offspring something to cling onto, the female *aquaticus* retained the hair on her head and grew larger breasts than her earthbound cousins.

The book had a major impact on its lay readers, but incurred the wrath of professional anthropologists and paleontologists, who decried its defiantly non-scientific approach. Morgan pressed on however, refining her ideas, and her language, in follow-ups including *The Aquatic Ape* (1982) and *The Scars of Evolution* (1990).

While undoubtedly appealing on many levels —after all who doesn't like dolphins and seals (other than Canadians)—the AAH has been repeatedly rejected by the evolutionary establishment. Amongst other things they point to the aforementioned lack of fossil evidence, the lack of bipedalism in any existing aquatic mammals and, contrary to the AAH, the fact that many marine mammals—e.g. seals, otters, polar bears—are covered in thick fur, which acts as a good insulator against cold water.

Morgan has held her ground however and retains a loyal base of supporters. Just when you thought it was safe to go back into the water, this is one marine battle that's not over yet...

020 THE MARS EFFECT

nown at school as Nostradamus for his prodigious ability to draw up astrological birth charts, the French psychologist and statistician Michel Gauquelin would dedicate much of his career to the study of the heavenly art.

Often in partnership with his wife Françoise, Gauquelin conducted several investigations into astrology's claims. His "test of opposed destinies" asked astrologers to study 40 birth charts and separate those of 20 criminals from 20 responsible citizens. Their results were in line with chance.

In another experiment he asked an astrologer to cast a horoscope for a birth date he claimed to be his own. The reading was extremely flattering: "He is a Virgo, instinctive warmth is allied with intellect and wit... He is endowed with a moral sense which is comforting—that of a worthy, right-thinking citizen... [whose] life finds expression in total devotion to others..." The birth date supplied was actually that of Marcel Petiot, a French doctor who had robbed and killed 27 people as they sought shelter from the Nazis during World War II.

He then sent this description, along with a questionnaire, to 150 respondents to a newspaper advertisement offering free astrological readings. Perhaps unsurprisingly, 94 percent declared themselves satisfied and impressed with his assessment of their personalities.

So Gauquelin must have been as surprised as anyone when one of his surveys, published in the late '50s, appeared to contradict his other, negative findings. Examining the birth dates of over 2000 prominent Frenchmen, Gauquelin found that certain planets appeared prominently in the birth charts of specific professions. Most famously, a clear link was made between the planet Mars and the birth times of sporting heroes. Gauquelin called his work "astrobiology," and this particular finding became known as "the Mars effect."

The experiments proved controversial to say the least, and haunted Gauquelin for the rest of his life. Over the next 30 years, they would be re-assessed and repeated several times by both advocates and opponents, but confusion reigned high and the follow-up results were nothing if not inconclusive.

Gauquelin himself seemed clear on the matter of astrology, writing in 1969: "The signs in the sky which presided over our births have no power whatever to decide our fates [or] to affect our hereditary characteristics." Sadly, the Mars effect had a devastating impact on its discoverer. Following a nervous breakdown, Gauquelin ordered all his files destroyed and committed suicide in May 1991, at the age of 60. ♂

021 PLANT SENTIENCE

Cleve Backster, was, and still is, highly respected in the now-controversial field of polygraphy, the use of lie detectors, and has worked for the CIA amongst other organizations. He is most famous, however, as the man who got us all talking to our plants.

In February 1966 Backster bought a *dracaena* cane plant from a neighbor and placed it in his polygraph room. Curious to measure how long it would take water to reach the end of its leaves, Backster decided to attach polygraph electrodes to the plant. The device's electrodes measure galvanic skin response—its electrical conductivity—and so they should, he thought, register a change when water reached the end of the leaf. What he noticed, however, was that the plant was showing readings similar to those that a human would.

Backster wondered if he could get the plant to have any other effect on the polygraph and tried dipping a leaf in warm coffee. Nothing happened. Then he considered burning the leaf with the result that the polygraph, claims Backster, "went wild. The pen jumped right off the top of the chart." Fetching some matches didn't seem to alter the already disturbed readout, but taking them into another room seemed to have a calming affect, and the polygraph reading returned to normal. The plant, it seemed, had registered stress.

Following this dramatic incident, Backster decided to investigate further, devising an experiment in which, in the presence of a plant, brine shrimp were dropped into boiling water at random intervals. It appeared that the plant was registering the shrimps' stress as they were plucked from safety then boiled to death. Could the plants be demonstrating not just some kind of sentience, but telepathy too?

Backster certainly thought so and named the idea "primary perception." In 1968 his work was written up in the *International Journal of Parapsychology*, and featured heavily in Peter Thomkins and

ELECTRIC LIE DETECTOR
.. A NEW FUN-PROVOKING STUNT FOR PARTIES

WITH this homemade electric lie detector, you can myself and entertain your friends by exposing falsehoods. Hundreds of tests by the writer have proved its meter to indicate correctly about eighty percent of the time, and that is sufficient to get a lot of laughs at any party.

The cost need not exceed $2.50. A three-element radio tube (No. 45) is used in the circuit. The other parts required are a socket, 45-volt "B" battery with a 22½-volt

tap, two 1.5-volt dry cells connected in series, two cheap laboratory test pods, and an inexpensive milliammeter. A 0-15 milliammeter was employed for the detector above, but one with higher readings is preferable because the readings are often near the 15-milliampere mark. A small rheostat can be placed in series with the milliammeter to reduce the readings, if preferred.

The parts may be mounted in any convenient way provided the connections are made as shown. It pays, however, to design the box or cabinet so as to look as impressive as possible for its psychological effect. The 45 volts applied to the plate of the tube through the milliammeter are enough to produce a current reading of about 15 milliamperes. By holding one of the test pods in each hand, the subject connects the 22½-volt tap of the battery to the grid of the tube. As the pressure of the hands upon the test pods increases, the resistance a

GIANT HOME WORKSHOP MANUAL

Christopher Bird's hugely popular book *The Secret Life of Plants* (1973). Soviet scientists also took an interest in his work, inviting him to chair a panel at the first Psychotronic Association conference in Prague.

Encouraged by this interest, Backster began to experiment further, wiring up other substances including yogurt bacteria, eggs, even human sperm. The results seemed to demonstrate that "primary perception" could be measured in all living objects, and that distance was no object. When one subject left some blood cells in Backster's San Diego lab, then flew 300 miles to Phoenix, Arizona, they still appeared to reflect his mood changes.

Naturally, Backster's findings are not without their critics. As with so many parapsychology experiments, repeatability is a constant problem—his results, and those of others who have tried the experiments, seem to be spontaneous, and refuse to comply with the usual rules of science. Some have criticized his lack of control experiments, others have suggested that the polygraphs are merely responding to the build up of static electricity in the room, or even Backster's own telekinetic abilities.

Whatever the case, Backster's seed of an idea has blossomed and flourished and is unlikely to disappear.

Digital technology has revolutionized photography, but 40 years ago a hard-drinking Chicago hotel porter, Ted Serios, demonstrated abilities that make today's hi-tech equipment seem primitive by comparison. For Serios claimed that, simply by concentrating his thoughts into a camera, he could imprint an image on to unexposed film.

A professional psychiatrist, Jule Eisenbud, outlined Serios's feats in *The World of Ted Serios: "Thoughtographic" Studies of an Extraordinary Mind* (1967), describing how Serios conjured up images to order, some of them from vantage points high above the ground.

Serios was not the first thoughtographer. Nikola Tesla was convinced that such a feat would one day be possible, while in 1910 Tomokichi Fukurai, a psychology professor at Tokyo University, conducted public experiments with Mifune Chizuko, an alleged psychic. One of his fascinations was with what he termed *nenshu*, or psychography—Serios's thoughtography. But Fukarai's demonstrations with Chizuko were considered a failure by the science establishment. The psychic was branded a fraud and the professor a dupe, leading to her suicide and his resignation, events that would later inspire the popular Japanese *Ring* cycle of films in the late 1990s. Fukarai continued his investigations, however, and in 1931 they were published in English as *Spirit and Mysterious World*.

While Ted Serios was making the news, a Russian psychic, Nina Kulagina, was under investigation by the military physiologist Dr. Genady Sergeyev. A famous film shows her moving small objects around on a table, apparently without touching them; she also demonstrated thoughtographic skills. Sergeyev noted unusual electrical activity in Kulagina's brain, as well as an increased pulse rate, which probably contributed to the heart attack that nearly killed her in 1970, aged 44.

Back in the US, Serios was running into problems. The October 1967 issue of *Popular Photography* ran an exposé by Charlie Reynolds and David Eisendrath, both photographers and amateur illusionists, who had spent a weekend studying the psychic celebrity. During this time, Serios could only produce thoughtographs with the aid of a small tube that he pointed at the camera, ostensibly to aid his concentration. Most skeptics believe that the tube was projecting an image into the camera lens and on to the film. Serios never fully recovered from this rude exposure and, in the following years, he and his powers gradually faded from view. He died, largely forgotten, in the early 1990s.

PART TWO

AMAZING INVENTIONS AND

SPIRITED INVENTORS

A
midst the ongoing chaos in Iraq, the fate of several thousand artifacts missing from the National Museum of Baghdad is likely to remain uncertain. Among them is an unassuming-looking, 13 cm (5.1 in) long clay jar that represents one of archeology's greatest puzzles—the Baghdad battery.

The enigmatic vessel was unearthed by the German archeologist Wilhelm König in 1938, either in the National Museum itself, or in a grave at Khujut Rabu, a Parthian (224 B.C.E.–226 C.E.) site near Baghdad (accounts differ). The corroded earthenware jar contained a copper cylinder, which itself encased an iron rod, all sealed with asphalt. König recognized it as a battery and identified several more specimens from fragments found in the region. He theorized that several batteries would have been strung together, to increase their output, and used to electroplate precious objects; but his ideas were rejected by his peers and, with the onset of World War II, subsequently forgotten.

Following the war, fresh analysis revealed signs of corrosion by an acidic substance, perhaps vinegar or wine. An American engineer, Willard Gray, filled a replica jar with grape juice and was able to generate 1.5–2 volts of electrical potential. Then, in the late 1970s, a German team used a string of replica batteries to successfully electroplate a thin layer of silver.

About a dozen such jars were held in Baghdad's National Museum. Although their exact age is uncertain, they're thought to date from the Sassanian period, approximately 225–640 C.E. While it's now largely accepted that the jars are indeed batteries, their purpose remains unknown. What were our ancestors doing with (admittedly small) electric charges, 1000 years before the first sparkings of our modern electrical age?

Certainly the batteries would have been highly-valued objects: several were needed to provide even a small amount of power. The electroplating theory remains a contender, while a medical function has also been suggested—the Ancient Greeks, for example, are known to have used electric eels to numb pain. Another possibility is that the jars were used in ritual circumstances, perhaps hidden inside religious statues to inject some real buzz into the gods. Others dismiss the tricky notion of ancient electricity altogether and suggest that the jars were simply containers—they are similar to ordinary storage vessels found nearby—and that their contents have simple rotted away in the ensuing centuries.

Unless other comparable objects are found by archeologists, the batteries are likely to keep their secrets—that is if they survive the devastations of this and future wars in the region.

"A more self-willed, self-satisfied, or self-deluded class of the community it would be impossible to imagine. They hope against hope, scorning all opposition with ridiculous vehemence, although centuries have not advanced them one step in the way of progress." *The Mechanic's Magazine* on perpetual motion enthusiasts, 1848.

The first known perpetual motion machine was a wheel self-powered by mercury running along its spokes, described by the eighth century Indian astronomer Bhaskara. Known as "over-balanced wheels," such devices recur throughout history; other contraptions have used water, atmospheric pressure, magnets, radiation, even gravity itself to beat the cycle of entropy. Several patents for perpetual motion machines were granted in the UK in 1635, most of them based on this same concept. Over 600 applications have followed and today both the UK and US patent offices refuse to grant perpetual motion patents.

Perhaps the only case not to be poo-pooed outright is that of Germany's Johann Bessler who, after making a detailed study of hundreds of perpetual motion plans, produced a device of his own in 1712. Calling himself Orffyreus and his construction Orffyreus'

Wheel, he convinced many scientists that he had at last solved this ancient problem. One of his devices is said to have continued to rotate for at least three months while being scrutinized by physicists. Even the Royal Society was debating whether to purchase one for themselves, at a staggering price of £20,000 (equivalent to about £2.5 million/$4.9 million today). But despite the interest of the scientific elite, Bessler vanished in 1718, taking his devices with him. In 1727 his maid, Anne Rosine Mauersbergerin declared that the machines were indeed a scam, being operated from another room, but was never able to provide evidence for her claims. Bessler himelf died in 1745, falling from a windmill in Furstenburg—whether it was a perpetual motion windmill remains unclear.

New sources of energy give rise to new hopes of perpetual success. The last device to be taken seriously was the radium clock, built by William Strutt in 1903, the year before he won the Nobel prize for the discovery of argon and became Lord Rayleigh. Inside an evacuated glass case, two gold leaves opened and closed

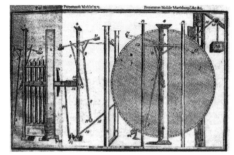

with no obvious source of power. Had Strutt cracked the problem? Many colleagues thought so, and it was some time before the mystery was solved. The leaves were actually charged by radioactive emissions from radium inside the container. It was a close shot, however: such a device could run happily for over a thousand years.

It's understandable that perpetual motion gets such short shrift from the scientific establishment: the very notion contravenes two of the laws of thermodynamics that govern our understanding of the physical universe. The first law states that "energy is conserved" or "you can't get something for nothing." Energy is never lost or gained in a physical process, it is converted from one form—in a perpetual motion machine usually kinetic energy—into another, such as heat, generated by friction. The second law concerns entropy. Its disorderly effects mean that, however cunning your perpetual motion creation, it is condemned to run out of steam eventually.

Contrary to what you might expect, physicists are quite grateful to the perpetual motion mechanics. These fundamental laws of thermodynamics were actually formulated in the nineteenth century to explain why no perpetual motion machines would ever work. The overbalanced wheel still displayed in London's Royal Institution is perhaps a tip of the hat to its wide-eyed inventors and suggests that this archetypal technological fantasy will always have a place in science's colorful history.

A fter studying medicine in his hometown of Edinburgh, Scotland, in the early 1770s James Graham headed to New England, eventually settling in Philadelphia as an eye specialist. Here he encountered the electrical demonstrations of Benjamin Franklin and realized that, with the right marketing brains, this strange new force could galvanize a veritable revolution in wellness.

In 1775 Graham shifted operations across the Atlantic again to London, making a name for himself providing electric alms for society ladies, thanks largely to his brother William, who was married to historian Catherine Macaulay. Clients, among them "it" girl Georgiana, Duchess of Devonshire, flocked to Graham to be treated with "aetherial balsams," electrical baths and a spell on his "magnetic throne."

As James' fame grew, so did his ideas, and in 1779 he opened the Temple *Aesculapio Sacrum,* known as The Temple of Health, on Adelphi Terrace, off London's Embankment. Equal parts health spa, museum and carnival attraction, this "magical edifice" boasted "magnificent electrical apparatus" and "supremely brilliant and Unique decorations." Graham himself acted as tour guide, "explaining the true Nature and Effects of Electricity, Air, Music and Magnetism when applied to the human body."

The temple offered a variety of educational and stimulating displays. As well as Graham's lectures, the curious could admire youthful, and beautiful, living specimens of perfect health, and try out various medico-electrical devices. And what museum would be complete without a gift shop? Here visitors could purchase blood-purifying Imperial Pills, enervating Aetherial Balsam, and cleansing Electrical Aethers.

The Temple's crackling, vibrating heart, however, lay in a private room on the second floor. For a whopping £50 (£5000/$9800 today), wealthy visitors could experience the ultimate sensual sensorium: The Celestial Bed. Here the Temple of Hymen was brought to dazzling life through music, perfume, art, sex and, of course, electricity. Bordered by huge mirrors, the inclining, magnetized, 6-½ ft (2 m) bed was charged with electrical current, creating an experience so potent that, even "the barren must certainly become more fruitful when they are powerfully agitated in the delights of love."

Sadly Graham's £10,000 investment failed to draw the necessary crowds, and by 1783 he had shut up shop and returned to Edinburgh. He ended his days lecturing on the benefits of fasting, healthy eating and mud baths, dying suddenly in 1794 from a burst blood vessel, at the age of 49. ⬭

"We are on the threshold of a gigantic revolution, based on the wireless transmission of power," wrote the Serbian electrical pioneer Nikola Tesla, then in his 70s, in the 1930s. "We will be enabled to illuminate the whole sky at night... eventually we will flash power in virtually unlimited amounts to (other) planets."

The modern world owes a great deal to Tesla's vision: alternating current, radio (he bested Marconi by at least three years), radio-controlled vehicles and fluorescent lighting are among his more mundane creations. Others remain highly esoteric, perhaps none more so than Universal Energy, in which power would be drawn from the air itself and supplied to homes, machinery and vehicles without wires or cables.

The trick, Tesla thought, was to use the air of the upper atmosphere to transmit energy—over any distance—above or straight through the Earth, even to other planets. Power would be beamed to a terminal floating in the upper atmosphere, then transmitted to any number of receivers on the ground or in the air. The risks were potentially high—"so strangely do such powerful discharges behave," he wrote in 1899, "that I have often experienced a fear that the atmosphere might be ignited."

In 1901 Tesla began to build a tower on America's East Coast that could transmit radio messages and, ultimately, power across the

Atlantic. Initial work at the Wardenclyffe site on Long Island, New York, produced spectacular lightning displays that were seen for miles around, but Tesla's ever-precarious financial situation meant that the tower was never completed.

Despite his reputation as a visionary and genius, Tesla's refusal to compromise his ideas saw a steady decline in his fortunes. His pronouncements about the future of radio communications, Universal Energy and robot technologies still commanded front-page news, but with successive wars and the Great Depression, his utopian dreams drew ever further from reality. Following his lonely death in 1943, many of the secretive inventor's papers vanished into the FBI's archives, leaving only speculation and mystery behind.

Although his more fanciful notions, like Universal Energy, seem unlikely ever to have succeeded, Tesla's lifework surrounds us. As the founding father of the alternating current, his legacy is felt every time we use an electrical device. But of all his creations, only the Tesla Coil bears his name. For aesthetic appeal, historic resonance and a satisfyingly high level of danger, the Tesla Coil is the ultimate fringe science icon, a role it has played to perfection in the laboratories of numerous Hollywood mad scientists.

The device is a step-up converter that uses two resonant coils to transform low voltage inputs from an ordinary plug socket into extremely high frequency, high voltage output. Usually channeled out via a toroid-shaped cap, the spectacular and intense discharge—ranging from a few centimeters to tens of meters in length, depending on the size of the coil—is essentially man-made lightning.

Tesla unveiled his first coils in 1891, to demonstrate his dream of wireless power transmission, which he would do successfully using a 49 ft (15 m)-wide coil topped by a 43 ft (13 m) mast at his Colorado Springs laboratory. This coil, which effortlessly generated millions of volts, reportedly produced great sparks 130 ft (40 m) long and energized the atmosphere surrounding the lab so that light bulbs outside could be lit wirelessly. It's said that the surrounding prairie shook as thunder and lightning erupted from the lab, while locals described sparks crackling underfoot some distance away.

Variations on Tesla's coil technology would be used for decades in radios, cathode ray tube televisions, and diathermic medical equipment. They're also used in decorative plasma sphere toys, which would no doubt please Tesla immensely.

In the 1970s artist and en-

gineer Robert Golka built a 121 ft (37 m)-tall Tesla Coil inside a hanger at Wendover Air Force Base in Utah. Golka, who declared his own coil to be more powerful than the one at Colorado Springs, initially attempted to create ball lightning with the tower, allegedly with some success. The coil was later commissioned by the US Air Force to test the effects of lightning strikes on aircraft, leading to rumors that the USAF was developing a lightning weapon.

Today Tesla's vision is kept alive by a global network of enthusiasts, known as coilers, who stage spectacular public demonstrations, hopefully inspiring future generations of lightning junkies. 🗗

027 SHOCKING HEALTH

" In recent years the use of electricity in medical treatment has made great strides...and it is recognized that a wide field of human ailments can be prevented, cured and relieved by suitable and timely electric treatment."

These reassuring words open a lengthy article in a late 1930s edition of *The Universal Home Doctor Illustrated*, detailing the many curative uses of electricity both in the hospital and the home. From hysteria to hemorrhoids, acne to asthma, bronchitis to bed-wetting, electricity—either via high frequency tingles on the

skin or shocking blasts to the central nervous system—was considered a near panacea, with some hospitals containing whole wards dedicated to its use.

Earlier, in the late nineteenth century, home users could purchase a dazzling array of electrical apparatus to cure them of just about anything. Dr. Carter Moffat's Electric Body Belt was effective against "liver torpidity" and melancholia, while J. Moses' Electro Galvanic Spectacles promised to "restore the organ of sight to its original strength."

Electricity was still a novelty at the time and this kind of equipment was expensive, but by the 1910s a new range of high frequency "Violet Ray" devices were being mass-produced for home use. These often came in attractive carry cases containing an induction coil, based on a Nikola Tesla design, attached to a handle into which glass vacuum tubes were slotted. These were shaped according to their intended use, and were usually filled with argon, which glowed purple when charged—hence the name. A standard kit might contain a flat-ended tube for use on the surface of the skin, a comb for the hair and head, and a tear-drop-shaped bulb for sensitive internal use.

Small Violet Ray devices were commonly found in the home, though the Depression saw a major drop in production. By World War II most manufacturers had converted their factories to making radios and other more useful electrical equipment. In 1951 the American Food and Drug Administration put a stop to cure-all claims associated with the devices, though they continued to be sold for use in beauty treatment.

Although little discussed at the time, Violet Rays were undoubtedly used to relieve tensions of a sexual nature. It's largely thanks to contemporary sensual adventurers on the fetish and bondage scenes that a specialist market exists for them today, with working models, usually made in Eastern Europe, being highly sought-after collectors' items.

028 KEELY'S TRICKSTER ENGINE

Born in 1827, John Ernst Worrell Keely lost both his parents as an infant. He is said to have left school at age 12 and spent time as both an orchestra leader and a circus performer. At 45 he was a carpenter in his native Philadelphia and had been for some years; but within months he would be a celebrated public figure.

In 1872 Keely announced the discovery of a new physical force, described, in typically vague language, as resulting from the intermolecular vibrations of the aether. If he could only build a machine to harness this force he would be able to produce vast amounts of energy by vibrating molecules at the atomic level. A year later he was being financially supported by a string of corporate and private investors, all hoping for a piece of the aetheric action.

The investors incorporated the Keely Motor Company and raised $10,000 for the inventor, who built a device he would name variously the "liberator," the "vibratory generator" and the "hydro-pneumatic-pulsating-vacu-engine"—but how it worked remained a mystery. If he remained vague about its mechanics, Keely was more than happy to show his machine off to potential investors and curious journalists. Even the British government sent a team over to investigate. Witnesses claim to have seen iron bars twisted or snapped in two, thick ropes shredded and bullets shot through 12-inch planks, all through no discernable means.

Keely was to be perpetually frustrated by his own inability to fully understand and exploit his creation, and he was never awarded the 30 or so patents that he needed to make it his own. This lack of demonstrable progress took its toll, and by 1880 his investors, now numbering 3000, were losing patience. Demanding answers they took Keely to court, prompting him to take a sledgehammer to his equipment in anger. His funding now came from a single source, Mrs. Clara Bloomfield-Moore, but even this had dried up by 1895.

Proportionate to his desperation, Keely's claims for his engine grew ever grander: a quarter gallon of water would power a train half way across the United States and a whole gallon could drive a steamer from New York to Liverpool; while a special belt would grant its wearers powers of levitation.

Afflicted by Bright's disease, Keely died on November 18, 1898. The Keely Motor Company arranged for an inspection of his workshop, and of course, his mysterious invention. The meaning of what they found is still hotly debated today. In a secret room beneath the engine was a metallic sphere, connected to the device by hollow brass tubes. His detractors identified it as a hydraulic device—if they were correct, then Keely's magical engine ran on nothing more mysterious than compressed air. The machine itself was transported to Boston, but never worked again.

Aetheric visionary, energetic charlatan or, perhaps, a victim of his own hype, Keely took his secrets to the grave.

Vibratory Planetary Globe
with wave plate, fork and spirophone.

"**A** flaming death… an invisible, inevitable sword of heat." So H. G. Wells described the Martians' heat ray in his 1898 classic *The War of the Worlds,* one of the first novels to introduce the death ray into popular consciousness.

Historical accounts of this archetypal weapon of mass destruction date back to the Second Punic War of 218–202 B.C.E. Defending the city of Syracuse against a Roman assault, the Greek sage Archimedes is said to have used a series of hexagonal mirrors (or bronze shields in some accounts) to focus sunlight into a searing beam that set the Roman ships alight. The incident was successfully reconstructed in 1973 by Dr. Ioannis Sakkas, who used 50 bronze-coated mirrors as reflectors.

When it comes to twentieth century death rays, two names feature prominently: the Serbian inventor Nikola Tesla and an Englishman, Harry Grindell Matthews.

"It will be possible to destroy anything approaching within 200 miles. My invention will provide a wall of power," the 81-year-old Tesla told *Liberty* magazine in 1937. Tesla spoke of several directional light-ning and beam weapons in his lifetime, none of which were publicly demonstrated. In his final years he announced something called "teleforce," a highly charged particle wall that would surround nations and destroy anything that approached it. Like many of the grand claims made by Tesla later in life, that of teleforce can probably be taken with a pinch of salt; nevertheless it's likely to have been one of the reasons that the FBI took possession of Tesla's papers after his death.

Over the Atlantic in England, Harry Grindell Matthews, who produced the world's first talking film in 1921, of Shackleton's speech before setting off for Antarctica, developed a "death ray" of his own. Matthews astounded journalists in 1921 by stopping a motorcycle engine at 50 ft (15 m) using a projected energy beam. The finished version, he claimed, would knock airplanes out of the sky and detonate ammunition dumps from great distances. Matthews refused to explain how the device worked, however, and a demonstration for the military proved unspectacular, only causing a light bulb to glow and a small motor to stop.

When the British government refused to fund his project, Matthews threatened to take it to the French, then the Americans. Neither nation produced working death rays during World War II, so we can assume that he failed to convince them too. A 1924 Pathé film features Matthews himself operating an impressive, but unfortunately entirely fictional, death ray cannon; probably as close to the real thing as he ever got.

For many years the small town of Murray, Kentucky liked to consider itself "The Birthplace of Radio." Its claim was based on the work of Nathan B. Stubblefield, a local mystic, inventor and melon farmer who, according to his proponents, beat Marconi to the first radio transmission (though of course Tesla beat Marconi too!) by some three years.

Among Stubblefield's numerous inventions was an earth battery that drew power from the ground and allegedly ran a motor for two months, but it was communication that really inspired him. Alexander Graham Bell's telephone exchanges were connecting big cities across the nation but were too expensive for small towns like Murray, so Stubblefield developed his own, using the ground itself as his medium. "My medium is everywhere," he would say.

From 1892 Stubblefield worked on a "wireless telephone" which used the ground's natural conductivity to send and receive transmissions. It apparently reproduced sounds significantly more clearly than Bell's cable telephones, but over shorter distances, managing a still impressive 1-¼ miles (2 km)—plenty far enough for Murray. A public demonstration on January 1, 1902 led to further requests from the cities of Philadelphia (where Tesla himself was in attendance),

New York and Washington, DC, where he used the telephone to communicate with offshore ships.

Following these demonstrations, New York businessmen bought out his patent and founded the Wireless Telephone Company of America. But Stubblefield soon became suspicious of his business partners and retreated to Murray where he tried, unsuccessfully, to patent a new system. As his financial situation worsened, Stubblefield's family left him to become increasingly eccentric and reclusive while he dreamt up new inventions. "I've lived 50 years before my time," he told one visitor later in life. "The past is nothing. I have perfected now the greatest invention the world has ever known." Sadly the world never saw this particular creation, or even found out what it was.

Stubblefield lived and worked alone until 1928, when he was found in his cabin, age 68, starved to death, his eyes eaten out by cats. The *New York Times* saw fit to publish his obituary and soon afterwards the legend of the man who invented radio was born, bringing much needed investment to Murray. Investigating the town's claims in the 1970s, researchers found little connection between Stubblefield's and Marconi's technologies. But, once born, legends never die—and Stubblefield's lives on. 🗘

031 SCIENCE IN CHAINS

America's prisons are traditionally a source of cheap labor, churning out blue jeans, license plates, glasses, cardboard boxes and similarly bland but practical items. But, according to a 1912 *Technology World* magazine article, one prisoner's engineering project at Arizona State Penitentiary so impressed his keepers that he was granted 30 days unaccompanied parole to patent it.

Roy J. Meyers was serving three and a half years at the time, in the spring of 1912, though the article doesn't mention what for. Prior to his conviction, he had applied for a number of patents, including one for a tram wheel, but his new idea was something altogether grander. Called the "Absorber," Meyers' invention would, he claimed, draw energy from the very atmosphere and convert it to electricity.

The prison authorities, perhaps hoping they had another Edison or Tesla on their hands, gave Meyers the use of a small workshop, rather surprisingly located just outside the prison walls. Within a few weeks, a demonstration model was igniting gas engines at the prison pump house, followed by a second device, which produced a modest output of eight volts.

A visiting prison reformer, Kate Barnard, suggested that Meyers be granted parole to pursue his work, and took his case to the state governor, George Hunt. Persuaded that Meyers could be trusted, Hunt gave him 30 days leave to bring his project to fruition.

Meyers' first stop was the Patent Office in Washington, DC. Already used to grandiose claims from would-be free energy pioneers, they suggested that he return with a working model. This he did, and so filed an application. Details are scant, but a photograph of a six foot model shows a wooden tower, atop which sits the Absorber itself: a ring of specially coated, magnetized steel plates connected to a transformer below. How it worked was, and still is, a secret.

Meyers returned to Florence to serve the last 10 months of his sentence, two days before his allotted free time was up. On his release, he told the *Technology World* article's author, he planned to build a 200 ft Absorber that would provide enough direct current to power the city of Phoenix. And that is where this tantalizing story ends.

Did Mr. Meyers ever build his full-size Absorber, or was he just another con man or crackpot? Barring a trip to Florence, Arizona, and then the US Patent Office, we may never find out. ⌒

t's said that future wars will be fought over water, not oil. One man who would have appreciated this statement was Viktor Schauberger, the Water Wizard.

Born in Austria, 1885, to a family who had been foresting for 400 years, Viktor grew up surrounded by unspoiled wilderness. Developing an intuitive understanding for the rhythms of nature, he became particularly fascinated by water, which he felt was a living organism as complex as any other. "Comprehend and copy Nature!" was his life's maxim.

Viktor first demonstrated his talents designing elaborate, highly efficient log flumes inspired, he said, by watching the movement of a snake through water. The unusually shallow, narrow flumes transported vast numbers of logs—including heavy oaks and beeches—through the valleys of Austria, Bavaria and Yugoslavia, and made him internationally admired.

The sight of a trout leaping from a fast-flowing river set the course for the next phase of Viktor's life. The trout, he observed, exploited natural forces and currents to propel itself powerfully out of the water. This led him to devise his own system of hydrodynamics, based on the spiraling motion of whirlpools and, later, wind vortices.

Following these observations Viktor next began to work on a radical new kind of engine that reversed conventional engineering practice, being driven by implosions, not explosions.

From here, Viktor's story becomes somewhat vague. The popular account has him being summoned to the court of Adolf Hitler in 1934, then, during World War II, blackmailed into working for the Germans, for whom he was forced to develop experimental air turbine engines, which were built into disc-shaped aircraft. These, according to some alternative historians, became the flying saucers that buzzed 1940s America.

After the war, an ageing and impoverished Viktor concentrated on techniques for improving agricultural yields and—his *magnum opus* —a home energy generator known as The Trout Turbine, based on his implosion theories. This is said to have attracted American entrepreneurs who, in 1958, took him—along with his plans, models and prototypes—to Texas. A proposed deal turned sour, and the strain was too much for Viktor, who returned to Austria, empty-handed, broken and exhausted. Schauberger died five days later, leaving behind an untapped legacy of ideas and very little paperwork.

" I use it every day," John Travolta told Germany's *STERN* magazine in 1997, "I'm always totally refreshed by it." The actor is describing the Hubbard Electrometer, Electropsychometer or E-meter, the device at the heart of Scientology's retro-futurist hybrid religion/therapeutic system.

Essentially a Wheatstone Bridge, as developed by Samuel Hunter Christie and Sir Charles Wheatstone in the first half of the nineteenth century, the Electropsychometer measures electrical resistance. In this case it is Galvanic Skin Response (GSR)—how sweaty one's hands are, and how well they will therefore conduct electricity. GSR is an important factor in the lie detectors, or polygraphs, developed in the United States during the 1930s and still used, controversially, by law enforcement and government authorities in several US states. Conversely polygraph evidence will not be accepted in most European courts of law as the procedure is deemed inaccurate and too easy to deceive.

In 1966 Scientology founder Lafayette Ron Hubbard was awarded a US Patent for a "Device for Measuring and Indicating Changes in Resistance of a Living Body," but the original Electropsychometer had been developed in the early 1950s by psychoanalyst Volney G. Mathison. Hubbard had adopted Mathison's device in the early practice of Dianetics, but when Mathison refused to relinquish the patent rights to L. Ron, it was dropped from the Scientology repertoire. It returned in 1958, however, after a smaller, more efficient version was developed by Scientology-friendly engineers.

Following a 1963 US Food and Drug Administration edict, Scientology can no longer refer to E-Meters as having any medical use. They are now "religious artifacts." Their own publicity states that the E-Meter "in itself does nothing," but the devices continue to play a central role in Church practice. The latest Mark VII Quantum model looks like two tin cans wired up to a Smart Car dashboard but it, like its predecessors, is integral to the process of becoming "Clear"—free from negative thoughts and emotions—a crucial goal for every Scientologist.

It works like this. The subject, or "Pre-Clear" (PC), holds a cylinder in each hand and is questioned by an "Auditor." As it moves through the body, the device's electrical current is, according to Scientology's instructions, influenced by the PC's thoughts and emotions. "The pictures in the mind contain energy and mass" and these generate electrical resistance (and, presumably, emotional resistance), which is registered by the E-Meter. The ticks and twitches of its dials are interpreted as evidence of negative energies that can only be overcome by signing up for a series of lengthy and expensive Scientology courses. As with lie detectors, cheating is relatively

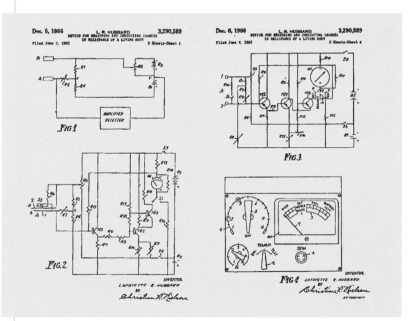

easy: relaxing reduces the needles' movement, while squeezing the cylinders induces sudden twitches.

Whether you consider Scientology to be sacred or profane, the Electropsychometer itself is innocent, and remains a fascinating technological—and, yes, religious—artifact. ▢

escribed in 1957 by arch-debunker Martin Gardner as promising to be "even funnier than Dianetics," engineer Thomas Galen Hieronymous's Type One Psionics Machine is an archetypical piece of magical technology.

A variant of the radionics devices created by Albert Abrams earlier in the century, the Hieronymous machine came to public attention through the enthusiasm of John W. Campbell Jr, editor of *Astounding Science Fiction*. Just to put things in perspective, Campbell also gave L. Ron Hubbard's "Dianetics" its first public platform.

The Hieronymous Machine has the unusual distinction of being the only psychically-operated—or psychotronic—device to have received a US Patent. On September 27, 1949 it was awarded patent 2,482,773 for the "detection of emanations from materials and measurement of the volumes thereof." Hieronymous claimed that his invention could detect "Eloptic Radiation," a form of energy emitted by "everything in our material world," which is then "tuned in" to by a psychically sensitive human operator.

According to Hieronymous everything, whether animal, vegetable or mineral, has an eloptic signature and vibrates at its own unique frequency. The component elements of more complex objects, like

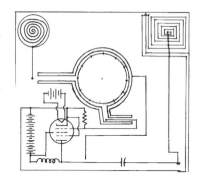

human beings, interact eloptically to create more complex signatures: in this way people can be identified from drops of sweat or blood. Because eloptic energy is conducted along light rays, photographs also contain eloptic traces of their subjects. The device's potential applications were almost unlimited: Hieronymous suggested chemical analysis, military photographic intelligence (which he claimed to have done—Top Secret, of course), horticulture and animal husbandry, amongst others.

Hieronymous later developed an eloptic transmitter, which he claimed could remotely heal living beings and, in one experiment, transmitted sustenance to plants in a lightless room. Like other radionics devices, the transmitter also proved useful in pest control: a photograph of a field or garden, and a sample bug, was enough to clear it of parasitic critters.

A radio ham and electrical tinkerer, John W. Campbell built a Hieronymous device himself, expecting it to be a joke; but to his surprise, it worked. In a truly bizarre next step, he discovered that not only did the machine itself—composed of electrodes, condensers, prisms and dials—work, but so did the printed circuit diagram, on its own. Yes you did read that right.

Fondly remembered by many *Astounding!* readers, the Hieronymous machine has, perhaps understandably, been forgotten by science. However, plans are available online (and reprinted here) should anyone care to put Campbell's wilder claims to the test. ♋

035 DR. RIFE'S OSCILLATING BEAM RAY

By the early 1930s, Dr. Royal Raymond Rife, an American optics engineer, claimed to be achieving theoretically impossible optical magnifications of over 30,000 times—ten times more powerful than today's best microscopes. At the time, only bulky and potentially dangerous, electron microscopes could reach comparable levels.

Soon after, Rife announced that he could destroy bacteria by blasting them with electromagnetic waves, oscillating at frequencies specific to each target organism. According to his supporters, Rife successfully cured significant numbers of people infected with a number of common but dangerous infections, including typhoid, salmonella and influenza. But his most controversial claim was that his device could kill the virus-like organisms, which he dubbed

"BX," responsible for cancer. Rife and his team claimed to have cured 15 "hopeless" cancer patients after just 60 days' treatment.

Rife's Raytube System was successfully installed in several clinics and numerous scientists and doctors corroborated his results. In 1939 he was invited to address the Royal Society of Medicine in London, which had also approved his findings, and he subsequently formed the Rife Ray Beam Tube Corporation to build models of his device for use in hospitals and clinics.

But with the death of one of his key supporters, Rife found himself under sudden and prolonged assault by the American Medical Association, who banned practitioners from using his Beam Ray in the treatment of patients. Within a year the dream was over, Rife a broken man. To this day it remains unclear why the AMA turned on Rife so brutally, a pharmaceutical conspiracy (usually centered around its then head Dr. Morris Fishein) being an obvious, if necessarily paranoid answer.

Rife died in 1971, but he never renounced his discoveries. Today he remains a hero of the fringe science underground, and

blueprints for his microscopes and beam rays are highly prized. In 2004 an English group claimed to have found an intact 1939 Beam Ray walled up in a doctor's office, which they have now reverse-engineered to a functioning level. The group remains cautious and is making no medical claims for the device, though they intend to replicate Rife's experiments. Other remnants survive in the archives of London's Science Museum.

Meanwhile, mainstream scientists are again examining the links between certain viruses and cancers. The papilloma virus, for example, is known to cause cervical cancer, while breast cancer has been linked to an HIV-like virus. After years in the scientific hinterlands, Rife's ideas may yet be reconsidered. Today FDA-approved, Rife-like devices use frequencies to purify water, sterilize food and treat arthritis.

"Part of the problem has been that the medical establishment does not have access to optical microscopes that can view living organisms at a high enough magnification," says Peter Walker, a leading Rife researcher. Walker is a co-founder of Grayfield Optical, whose Ergonom 500 Grayfield microscope can achieve 25,000x magnification with living samples, a technological advance he hopes will prove Rife to have been right all along. ▢

"**H**is gestures, his intonation; how powerful they were! What flights of oratory!" writes Father Pellegrino Ernetti, describing a speech by Marcus Tullius Cicero to the Roman Senate in 63 B.C.E. Ernetti, who died in 1992, was a Benedictine monk, a respected historian of ancient music, author, physicist and exorcist; but what makes his description of the orator's performance intriguing is that Ernetti claims to have witnessed it first-hand.

Well, sort of. Ernetti actually told the French theologian Father Francois Brune that he watched it live on a time-warping device he called the Chronovisor. Rather than transporting people through time, the Chronovisor tuned into the events of the past and displayed them like a time-traveling television.

Ernetti claimed to have developed the Chronovisor during the 1950s, along with 12 famed scientists who, except for Enrico Fermi and Wernher Von Braun (both conveniently deceased when their names were published in 1992) wished to remain anonymous. It consisted of numerous antennae, three composed of "mysterious" metals, that received light and sound signals on every wavelength; a "direction finder" for tuning in to a particular time and place; a screen and a recording device.

Ernetti described how, crowded around their invention, the time team watched speeches by Mussolini and Napoleon and scenes from ancient Rome, including a performance of Thyestes, a lost play by Quintus Ennius, one of the Roman Republic's most important authors.

Although nobody saw the Chronovisor itself, Ernietti did provide evidence of these televisual travels, for example publishing fragments of Thyestes in Latin. Then, on May 2, 1972, the Italian magazine La Domenica del Corriere published what should have been Ernetti's trump card: a Chronovisor image of the crucified Christ. The team had viewed his last days in their entirety: from the Last Supper to his final agonizing moments, and had retained the photo as a powerful souvenir.

But Ernetti's glory was short lived. A few months later another magazine revealed that the image of Christ was a reversed image of a postcard from the Santuario dell'Amore Misericordioso, a chapel in the Umbria town of Collevalenza. More recently, serious doubts have been cast on his "transcription" of Thyestes, and what appears to be a deathbed confession has surfaced. Perhaps it's time to add holy hoaxer to Ernetti's impressive list of accomplishments.

"**B**ooks are read while one sleeps. Most of the studying is done while one sleeps. Some people have mastered 10 languages, which they have learned during their sleep-life." *Modern Electronics*, June 1911.

Hugo Gernsback, founder of the first science fiction magazine, *Amazing Stories,* dreamt up the "hypnobioscope," a sleep learning device, for his popular "scientifiction" magazine column *Ralph 124C 41+: A Romance of the Year 2660.* The device read information taken from another person's brain (using what now sounds like an EEG machine), and beamed it directly into that of a sleeping subject.

A book version of *Ralph* appeared in 1925, and by 1927 the Psycho-Phone had been trademarked: users recorded a message onto a wax cylinder then, using a timer, set it to play repeatedly while they slept. Whether the Psycho-Phone was directly inspired by the hypnobioscope is unknown, but it was the closest thing to Gernsback's model that technology allowed.

In the 1940s and early '50s, experimenters demonstrated that the use of repetitive phrases during sleep could alter human behavior. Dubbed hypnopaedia, perhaps ironically, after a process described in

Aldous Huxley's *Brave New World* (1932), the experiments helped sleeping subjects to learn languages and break bad habits.

"Tomorrow's university will be at the bedside" promised the Linguaphone Company's 1948 advertisement for its Cerebrophone: "a revolutionary way to learn a foreign language while you sleep." Basically a phonograph connected to a timer and an under-pillow speaker, the device never caught on with the public, but it was to play a part in one of the more sinister chapters of American history.

In 1953 the Canadian psychiatrist Dr. Donald Ewen Cameron began to use the Cerebrophone, renamed the Dormaphone, as part of a technique he called "psychic driving." Following severe electroshock therapy, during which "unwanted" memories were erased, Cameron's exhausted patients would fall into a deep sleep, sometimes for several days, while messages were played to them repeatedly through the Dormaphone. Cameron, notorious for his cold and often unethical approach to patients, became a key figure in the CIA's MK-ULTRA investigation into brainwashing and mind control, where psychic

driving was used to "reprogram" a number of subjects. Some later sued Cameron for malpractice.

Sleep learning remains a popular tool to this day, with a choice of programs ranging from language learning to improving psychic abilities. Contemporary proponents reference several scientific studies into the efficacy of the procedure, but Cameron's work, which echoes Huxley's dystopian vision rather than Gernsback's brighter future is, understandably, invariably ignored.

038 THE NEUROPHONE

At eight years old Pat Flanagan was plagued by a recurring dream. He was an adult, flying a small airplane, when suddenly the engine died, forcing him to land on a small island. A strange aircraft then descended, out of which emerged beautiful blonde beings who measured his intelligence using a silver helmet. If his intelligence didn't match their expectations, the beings said, he, and the rest of his species, would be destroyed.

At 15, Flanagan had already begun to demonstrate the invention that would change his life—the Neurophone. Built in his home laboratory from wire and Brillo pads, the device transmitted audio signals from a stereo directly into the brain, bypassing the ears entirely.

Although he knew that the sounds were somehow being picked up by the wearer's skin and bone, the exact mechanism would evade the inventor for another 33 years.

Nine years after his dream, Flanagan was the subject of a September 1962 *Life* magazine profile that described the now 17-year-old inventor, pilot, gymnast and champion twister to be a "unique, mature and inquisitive scientist."

Following the *Life* article, Flanagan demonstrated his creation on television and in person, but the US Patent and Trademark Office remained unimpressed. It was only in 1968, after he successfully demonstrated it by playing opera to a deaf member of staff, that he was awarded patent number 3,393,279.

In his late twenties, Flanagan began training dolphins for the US military, teaching them to recognize the sonar signals of steel alloys used in ships and submarines. He and Dr. Wayne Bateau built a new Neurophone that translated 30 human words into dolphin clicks, a device that was so successful that the US Defense Intelligence Agency declared it a state secret for several years.

In 1991, the University of Virginia's Martin Lenhardt recreated Flanagan's findings using ultrasonic signals. He discovered that the saccule, a pea sized organ in the inner ear usually associated with balance, is also sensitive to ultrasonic sound, at last explaining how the Neurophone worked.

Flanagan, who became the leading proponent of "pyramid power" during the 1970s, now markets several Neurophones for relaxation and speed learning, as well as a powerful hydration supplement. It would seem he has no choice but to keep inventing—the Aryan aliens of his youth could return at any time.

039 ELECTROGRAVITICS: HUMAN FLYING SAUCERS

O n January 31, 2003, less than 24 hours before the demise of Space Shuttle *Columbia* and her crew, NASA announced that it would no longer be funding its Breakthrough Physics Propulsion Program (BPPP). This sounded the death knell for what was possibly the world's largest visible antigravity research project.

A working antigravity technology is still decades away, perhaps longer, but the origins of one of the most promising branches of aerospace research can be found in an Ohio garage, over a hundred years in the past.

Thomas Townsend Brown was born in Ohio in 1905, two years after the Wright Brothers achieved lift off. By his teens he was already dreaming about space travel, and while still a schoolboy he began tinkering in his parents' garage on the ideas that would obsess him for the rest of his life. One day, while experimenting with X Ray tubes, the teenager applied a high voltage electrical charge to a capacitor attached to a glass tube suspended from the ceiling. To Brown's astonishment, the tube began to rotate, apparently propelled by electricity itself.

At age 18, back in Ohio after an unsuccessful spell at the California Institute of Technology, Brown was taken under the wing of Dr. Paul Biefield, a close friend and colleague of Albert Einstein. Biefield was deeply impressed with Brown's discoveries and together they proposed the Biefield-Brown effect. This states that when electrical current is applied to a capacitor, it will move in the direction of the flow of current—towards its positive pole. And so the fledgling field of electrogravitics was born.

By the early '50s, Brown had developed a number of electrically-propelled, disc-shaped platforms, up to three feet in diameter, which he is said to have demonstrated hovering and rotating in a vacuum to impressed aeronautics scientists in the US, UK and France. The US military swiftly classified the results of Brown's demonstrations (and, according to some reports, set about building large scale versions of his discs) but they neglected to support his research. This situation would, sadly, prove to be the story of his life—Brown never got the funding he so badly needed, though he continued to develop his ideas until his death in 1985.

The story continues in the research of several contemporary groups exploring electrogravitics, among them Boeing, NASA and BAE Systems. Closest to Brown's original vision are the "Lifters" being built by American Antigravity and others—skeletal metal frames that can lift a pound in weight, propelled by ion winds generated by DC electric currents. Impressive as this is, it is not a true antigravity effect—the Lifters won't budge in a vacuum and so are no use to anyone in space. ▢

040 BODIES OF LIGHT: KIRLIAN IMAGERY

In 1939, while repairing high frequency electrotherapy equipment at his workshop in Krasnodar, Russia, the inventor and electrician Semyon Davidovich Kirlian made a spectacular discovery. When he attached a sparking electrode to his hand and placed it onto a photographic plate, the plate revealed the image of a glowing, blue, hand-shaped halo.

Over several years, Kirlian and his journalist wife Valentina developed equipment that allowed them to view moving electrified objects in real time, creating dazzling visual effects. Encouraged by visits from scientific dignitaries, the Kirlians became convinced that their bio-luminescent images showed a life force or energy field that reflected the physical and emotional states of living subjects and could even diagnose illnesses. In 1961 they published their first paper, in *The Russian Journal of Scientific and Applied Photography*, and their findings reached the West via the highly influential 1970 book *Psychic Discoveries Behind the Iron Curtain*.

The Kirlians' images struck a chord with UCLA parapsychologist and LSD-therapy enthusiast Dr. Thelma Moss. Moss made several trips to see them in the USSR, where their techniques were being taken seriously and used in space research. Particularly impressive were claims that, when a leaf is cut in two under Kirlian imaging, an outline of its missing half persists for some time, suggesting, they felt, the presence of an "energetic body." Moss considered the Kirlians' images documentary evidence of the subtle bodies and auras described by mystics for millennia, and used them to promote traditional Eastern ideas about healing and life force.

What the images actually reveal, as scientists at Philadelphia's Drexel University discovered in the late 1970s, is a kind of corona discharge—similar to that produced when dragging your feet on a carpet, or in St. Elmo's Fire. The team, partly sponsored by the US Defense Department, found the visual effect to be dependent on moisture levels on the subject's skin, similar to how a lie detector works, or the Scientologists' E-Meter. In a blow to the subtle body, the team failed to recreate the "phantom leaf" effect, attributing it to residue on the photographic plate, or even fraud.

Whatever it is that they show, the Kirlians' images are undeniably beautiful and have a rightfully iconic status in the history of photography, while a small industry of aura photographers, to be found at any New Age fair, combine their work with Thelma Moss' mystical notions. ⌑

041 THE POLTERGEIST MACHINE

"I'm certainly not the 'typical' scientist," says Vancouver's John Hutchison in a moment of monumental understatement. "I'm basically self-educated and don't use equipment manuals; nor do I take lab notes. I simply work as an artist does—with an intuitive feeling."

Inspired since childhood by his heroes Gene Roddenberry and Nikola Tesla, Hutchison has befuddled and frustrated physicists the world over with a series of bizarre physical phenomena collectively known as The Hutchison Effect. In his home-cum-laboratory, where most of the furniture has been removed to make room for all of his equipment—an astounding array of home-made Tesla coils, Van der Graaf generators, radio frequency generators and parts, including

radars salvaged from naval battleships—Hutchison claims to have twisted the laws of physics, and more besides.

It all began in 1979 when, having amassed an impressive hoard of equipment and fired by the spirit of invention, Hutchison decided to see what would happen if he turned everything on at once. With the kit humming away, he was stunned to see a bar of steel rise from the floor for an instant then drop back down with a bang. After a sleepless night, he tried it again, but nothing happened—whatever the phenomenon was, it would prove itself to be highly erratic.

More levitations followed over the ensuing months, and Hutchison began to film what was going on—glass bottles and tubes, tools, bits of metal and paper were seen shooting across the room and hovering off the ground. In a particularly surreal episode, yogurt rose slowly from a tub affixed to the ground. In time, far stranger things began to occur: hard alloy metals became as soft as putty, wood became embedded in once-solid metal and blocks of concrete suddenly burst into flame. Hutchison and his supporters stress that, whatever it is, the effect is not due to electrostatic or electromagnetic fields: it's something else. The problem is, nobody can work out what that something is.

As the levitation footage circulated, word soon reached the US military, who took one look and decided to pay Mr. Hutchison a friendly visit. Impressed by what they saw, they diverted enough money his way to keep him going and see what marvels he would produce next. In 1985 a team of military and aerospace scientists set themselves up at the lab in the hope of studying the Hutchison effect under controlled conditions. Nothing happened and the team lost interest.

Since then Hutchison has worked with scientists from Japan and Germany, though his relationship with the authorities has deteriorated quite dramatically: in 1990 his original lab was confiscated by the Canadian government, and in 2000 it was raided by armed police who, bizarrely, confiscated Hutchison's collection of antique weapons, then returned them a few days later.

The playful nature of the "Hutchison Effect" and the idiosyncratic nature of its discoverer have done neither any favors in the scientific community. The films themselves, while puzzling and often spectacular, are usually shot in close up—for example we see the yoghurt "rising" from its pot but have no idea whether the tub is being held upside down or the film has been reversed. This makes it extremely difficult to discount the possibility of fraud, though Hutchison himself has never been caught in the act.

A true "outsider scientist," Hutchison and his enigmatic machines look set to puzzle the rest of us for many years to come.

On March 23, 1989, two well-respected chemistry professors at the University of Utah, Martin Fleischmann and Stanley Pons, made an announcement that looked, for a brief moment, like it might change the world forever.

The professors claimed to have achieved nuclear fusion—normally produced by the intense heat and pressure inside stars—in a glass jar, at room temperature. This was "cold fusion." What's more, their fuel cell was pumping out four times the energy that was going into it. As the University of Utah's press release stated, the pair may have discovered an "inexhaustible source of energy"—the Holy Grail of physics.

The media went wild, predicting an end to fossil fuels and a fusion reactor in every home by the year 2000. But the excitement was short lived. The details provided by Fleischmann and Pons failed to satisfy their already hostile critics, who turned on the pair in a series of scathing attacks.

The problems were two-fold: they had a result without any theory to explain it and, to make things worse, initial attempts at recreating the experiment—by institutions like MIT (a world center of hot fusion research)—failed. Later investigations would bring MIT's own findings into disrepute, but the damage was done. Dodging accusa-

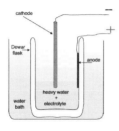

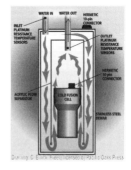

tions of fraud and incompetence, the pair fled the US and set up labs in France.

A few months after the cold fusion story broke, electrochemists at Texas A&M University repeated Fleischmann and Pons' experiment, finding small amounts of the radioactive by-product tritium in their jar and demonstrating that fusion had indeed taken place. Again, they were accused of fraud by science journalists, but no evidence for this was ever produced and the floodgates had been opened.

Like the hermetic alchemical networks of the seventeenth century, a worldwide cold fusion underground has sprung up in the aftermath of the original debacle, populated by respected researchers working quietly within academic institutions, and inevitably, not a few cranks. Since 1989 hundreds of papers have detailed successful replications and refinements of the original F-P experiment. There are enough researchers to meet for a large annual conference, regularly attended by several hundred people.

And we may be getting closer to an answer. A San Diego project set up by the US

Drawing: C. Erick Fleury, licensed by Pacific Oaks Press

Navy recently announced a potential breakthrough in our understanding of what's now referred to as the "Fleischmann-Pons Effect" or F-PE. The elements used must, they state, be of the utmost purity for the desired results to be achieved. In a further echo of chemistry's alchemical past, the lab claims to have witnessed the transmutation of metals into gold—traces of which bear testimony to the nuclear fusion process.

As our fossil fuels, and so our time, runs out, scientific luminaries including Arthur C. Clarke, have called for renewed vigor in researching this potentially planet-saving source of energy. 🖸

043 STARLITE

In 1993 viewers of the BBC's *Tomorrow's World* TV program were treated to a curious sight. Presenter Michael Rodd donned welding mask and gloves, fired up an oxyacetylene torch and used it to attack an apparently ordinary egg.

After four minutes the egg was cracked open to reveal its interior, still raw and runny. The secret? A thin coating of Starlite, a miraculous, heatproof substance invented by a middle-aged hairdresser from Blackburn, England.

Maurice Ward had no scientific background, though he once drove a forklift truck at an ICI warehouse. He did like to tinker however, and inside his rented workshop had created his own brand of hair products, using ordinary chemicals and a food mixer. He stumbled upon Starlite and its heatproof properties by accident and, realizing he had something special on his hands, began contacting chemical companies.

At first Ward was ignored as just another harebrained inventor, but following a brief mention in *Janes International Defence Review* and the *Tomorrow's World* demonstration, a number of defense and chemical companies took notice. More articles in intelligence and business journals followed, reporting that Starlite had withstood a simulated 10,000 °C (18,000 °F) nuclear blast and effortlessly shrugged off a high-powered laser beam.

Countless applications could be imagined for such a tough, lightweight material: from ship, aircraft and spacecraft design to body armor and kitchen equipment. Concerned, however, that a corporate giant would steal his secret recipe, Ward refused to patent his invention, revealing only that it contained 21 components, mostly organic polymers, borates and ceramics.

Rumors soon began to circulate: potential buyers included NASA and the US Department of Energy, both rejected when they refused to sign non-disclosure agreements; the plans had already been stolen and a secret factory was producing Starlite for the US government; other governments had tried unsuccessfully to replicate the mixture.

There was talk of billion dollar buyouts, stock market floatation, even a Maurice Ward Starlite Technology Center staffed by NASA employees. Then everything went quiet.

The last word came in 1997, when a spokesman announced that Ward was close to a major deal with a manufacturer of aircraft interiors. Maurice, meanwhile, was enjoying the thrills of his new hobby—harness horse racing.

A quick check has revealed that his horses were still running, with some success, in 2002. But whatever happened to Starlite? At the time of writing, Maurice Ward is back in action and now has a Starlite web site.

044 EXTREME NOISE TERROR

In 2003 the American Technology Corporation (ATC) revealed a new sound cannon for the Pentagon's "non-lethal" weapons program. Known as the Long Range Acoustic Device (LRAD), the unit directs painful or disturbing sounds towards a target using ultrasonic audio beams, and is said to be capable of disorientating or incapacitating a person at up to 500 meters. It was most famously—and successfully—used by a cruise ship to repel pirates off the coast of Somalia in 2005.

Sound's potential as a weapon has been understood since at least

1400 B.C.E. when, according to legend, the Israelites brought down the walls of Jericho by blasting them with ram's horns. In the late nineteenth century the Croatian electrical pioneer Nikola Tesla is said to have produced both intense pleasure and violent sickness in people using a vibrating infrasound chair, while during World War II, one Dr. Richard Wallauscheck is supposed to have built a Luftkanone or Sound Cannon in Germany. Burning methane and oxygen together, the Luftkanone produced an irritating tone that could be directed towards a target with a reflector. The device was estimated to be able to kill a human at 200 ft (60 m) in under a minute, though no such testing is known to have occurred.

Most famous of the post-war researchers is the late robotics engineer Dr. Vladimir Gavreau. While working in a huge, concrete research facility in Marseilles in the late 1950s, Gavreau and his team repeatedly became extremely nauseous. On investigating they found the cause to be a low frequency sound wave emanating from a faulty air-conditioning engine. Experimenting further with infrasound, the team are said to have almost ruptured their internal organs and destroyed their entire workshop. Gavreau is rumored to have constructed a series of potentially deadly infrasound whistles and horns, though no patents for these devices have been found.

More recent applications of sound weaponry include the "Squawk Box," allegedly used by the British Army in Northern Ireland in the 1970s, and "The Scream," touted by the Israeli army in 2005 as a peaceful weapon for use in crowd control. Eye-and-ear-witnesses to the latter say that it emitted a nausea inducing buzz at ten second intervals, which could be heard even with one's ears covered.

The key problem in creating sonic weapons is the direction and containment of the sound waves, but ATC claims to have solved it with its ultrasonic beams. The LRAD has already been deployed by police departments in the United States, including by the New York Police Department during the Republican National Convention protests in 2004, but we are more likely to encounter its commercial cousin, which is planned for use in emitting tightly focused sound-beams from soft drink vending machines. Now *that* sounds frightening. ⌨

045 INVISIBILITY: THE EMPIRE'S NEW CLOTHES

n 2005 the world's press carried photographs of a Japanese man wearing something that children everywhere, and not a few adults, have dreamed of owning—a cloak of invisibility. The image was created by Susumu Tachi, a Japanese professor of physics and computer science at Tokyo University, but was a mock-up made using good old-fashioned camera trickery. Images captured in real time by a camera behind the cloak were beamed onto it from a projector in front.

Accounts of invisibility have occurred in every culture, from vanishing medieval saints, through "The Invisibles" who announced their presence in seventeenth century Paris, to West African marabout priests who sell invisibility charms to this day. Cases of "spontaneous human invisibility" are also sometimes reported, whereby people find that for brief periods of time it seems that nobody can see them, leading to understandable anxiety and confusion.

Magical amulets won't get you very far in contemporary warfare, but the ability to vanish from your enemies has been a crucial advantage on the battlefield since the British first donned khakis in the late nineteenth century. In 1941, the English illusionist Jasper Maskelyne made the Suez Canal and the harbor of Alexandria "disappear" from enemy bombers. Also in World War II, America developed Project Yahootie, in which lights strung around the wings and fuselage of naval bombers were adjusted to the brightness of the sky above, making them difficult to spot with the naked eye. Although the technique proved more trouble than it was worth, these were the first stealth aircraft. The so-called "Philadelphia Experiment" of

1943, in which the American destroyer USS Eldridge vanished from sight in a naval yard is, unfortunately, folklore.

While invisible to radar, today's bulky stealth aircraft are easily spotted with the naked eye. But military technicians are said to be developing "adaptive camouflage" that changes color to match its surroundings, using a technique much like Susuma Tachi's. The idea is that sensors all over an aircraft receive visual information about their surroundings, then output corresponding images to chameleon-like panels made of electrochromic polymers on its bottom. Whatever is above the plane is projected onto its underside, making it virtually invisible from the ground.

Rumors persist within the world of secret aircraft and UFO enthusiasts that such technologies are already being used in next-generation vehicles being flown, ironically enough, at Area 51 in Nevada, the world's most photographed secret airbase.

046 ANGEL LIGHT

Canadian inventor Troy Hurtubise looks set to rewrite the physics rule book if his newest creation lives up to the claims of its supporters, who apparently include MIT engineers and French government officials. The Angel Light, they say, makes solid objects entirely transparent.

Hurtubise is no stranger to science headlines, having received a 1998 *Ig Nobel* prize in Safety Engineering for his *Robocop*-inspired bear-and-bulldozer-resistant suit. He has since developed a flame-and-heat-proof substance, Fire Paste, that withstands blowtorch temperatures and cools extremely quickly (highly reminiscent of Maurice Ward's Starlite) and, more recently, lightweight, blast-proof cushions. Any of these in themselves could prove revolutionary, but his newest creation could, literally, change the way physicists view the world around us.

The 8 ft (2.5 m) long Angel Light, looking like a sci-fi laser cannon, came to Troy in a dream, as, he says, do most of his inventions. "I saw it, the whole casing and everything, and I saw what it could do," he told a Canadian newspaper, "I had the same dream three times, and by the third time I had it in my head and I started to build it." The device is composed of three units: the "centrifuge" which includes seven industrial lasers and black, white, red and fluorescent light sources; the "deflector grid," containing optical glass, a microwave generator and plasma mixed with carbon dioxide; and a third part made of 108 mirrors, powerful lights and magnets, eight ionization cells and some trade secrets.

When shone onto a surface, the light temporarily creates a transparent window-like effect, allowing users to see straight through to the other side. It shines through wood, ceramics, plaster, metals (including steel, lead and titanium), and also skin. "I could see my blood vessels, muscles, everything, like I'd cut into my skin and peeled it back," said Hurtubise.

The cannon has some interesting side effects: it knocks out electrical devices including motors and televisions and cancels the effects of radar-resistant stealth technologies. It also has a detrimental effect on living creatures: until he fitted a shield, using it made Hurtubise seriously ill; several goldfish died quickly after being targeted in their bowl.

A host of detractors have denounced the invention as a hoax and Hurtubise as a con man, but if it works, and the military are interested as Troy has hinted, there's a strong chance that we may never hear of it until it's deployed on the battlefield.

As its creator might say: watch this space. 🖵

PART THREE

NATURAL MARVELS

The only undisputed twentieth century instance of science heresy becoming science orthodoxy is the case of continental drift.

Today its discovery is usually associated with the German meteorologist Alfred Wegener who, in his 1915 book *On the Origin of Continents and Oceans,* proposed that the Earth's continents were once a single landmass he named Pangaea. Until his death, Wegener's theory was regarded as preposterous, but he was not the first to make the claim.

In 1858 the French scientist Antonio Snider-Pelligrini suggested that the Atlantic Ocean had formed following the break-up of a single, huge continent. His ideas were rejected out of hand but, soon after, the Austrian geologist Eduard Suess proposed the existence of Gondwanaland, a vast landmass of continents connected by great land bridges, to explain the similarity of fossil plant finds across the Southern Hemisphere. In the 1870s, a British expedition on HMS Challenger charted the Atlantic Ocean ridge that separates Europe and Africa from America, but it was not until 1908 that the American geologist Frank Taylor would suggest that the continents had at one time been joined.

So although Wegener's Pangaea appeared to come out of nowhere, his ideas were firmly rooted in pre-existing knowledge, and his evidence was drawn from multiple disciplines, including geology, climatology, and paleontology. Nonetheless the response of the geological establishment to such a revolutionary theory, especially one coming from an outsider, was hostile in the extreme. Wegener was branded a heretic, his ideas dismissed by leading geologists as "geopoetry," and "utter, damned rot."

Wegener's Achilles Heel was his uncertainty over how the process of continental drift might actually take place. His initial ideas focused on the centrifugal force created by the spin of the Earth, or perhaps the gravitational pull of the Sun and Moon on our planet. In the fourth and final, 1929 edition of his book, Wegener conceded that he had not discovered a fully satisfactory mechanism, but that the same processes were responsible for oceans, mountain ranges, volcanoes and earthquakes. He died a year later, on an expedition to Greenland.

It was not until the 1960s, following advances in the mapping and dating of the ocean floor, that Wegener would be vindicated. Ironically, while continental drift is now part of establishment science, there is still no consensus as to the force behind it—the same force that had so eluded Wegener. The current hot favorite—convection currents in sub-surface magma—was posited by Wegener himself in 1929.

"The Newton of drift theory has not yet appeared," wrote Wegener in

the last edition of *On the Origin of Continents and Oceans*. Wegener
may not have been its Newton, but he might just be its Galileo. ◻

048 PET PREDICTORS

A round 3 million people have been killed during earthquakes
in the past century. Specialized construction techniques
have saved many thousands of lives in wealthier nations,
but 2003's quake in Bam, Southwest Iran, in which approximately
30,000 people died, demonstrated the full horror that these disasters can bring.

A number of esoteric indicators are said to precede earthquakes.
Psychic predictions by humans must rate as the least reliable, although claimants include Johann Goethe. In 1783 he had a vision
of a huge earthquake, the same night that 50,000 people were
killed in a quake in Calabria, Italy. Unusual luminous phenomena—glowing rocks, flashes and balls of light in the sky—have also
been recorded before large quakes, as well as low frequency electromagnetic pulses, erupting water and oil wells, gaseous ground
emissions and sudden changes in air temperature.

But the most controversial earthquake detectors are living creatures. Reports of pets, zoo and farm animals behaving strangely before quakes date back to ancient Greece. All manner of species,

from horses to bees, appear to be affected, sometimes seconds, sometimes weeks before an earthquake hits. It would seem likely that the actual environmental triggers for their behavior, from escaping gases to subtle vibrations, vary between species.

Following the South East Asian tsunami of 2004, reports circulated that animals including elephants, dogs and flamingos had been seen behaving strangely in the days preceding the devastating earthquake, some fleeing to higher ground. However, the stories remain anecdotal and have been largely dismissed by seismologists.

The most famous case of alleged animal activity prior to a major quake occurred in Haicheng, China in February 1975. Official Chinese statements claimed that by observing the behavior of animals in the city, authorities had been able to evacuate a million people, days before a 7.3 magnitude quake struck.

Following these reports, the US Geologic Survey commissioned a major study of the ways in which animal senses could be put to work saving human lives. Following a 1977 quake in California, researchers found that 50 percent of people surveyed near its epicenter had noted unusual behavior in their pets, but the team's findings weren't considered solid enough to justify any further action.

More disappointments followed. The reports from China were actually false—deliberate misinformation, perhaps intended to demonstrate that even the nation's animals were working towards the collective good. In fact, a number of smaller quakes had occurred in the Haicheng area, prompting the authorities to warn people to stay outdoors. This is what happened on the night of the big one. It was this very human observation, rather than quick-thinking critters, that had saved so many lives.

049 LIFE'S A GAS

Unusual natural phenomena are likely to account for a large number of so-called UFO encounters; this was the conclusion of the recently uncovered Project Condign report into UFOs, commissioned by the UK government between 1997 and 2000.

Amongst the most intriguing of these phenomena are "earth-lights," first properly studied in the mid-1970s. In the UK, Paul Devereux and Andrew York mapped out areas of UFO activity in Leicestershire, comparing them with maps of tectonic strain along geological fault lines. Two years later Michael Persinger and Gyslaine Lafrenière carried out a similar comparison of the entire USA. Both teams found striking correlations between areas of active tectonic pressure and accounts of unusual lights in the sky.

One of the mechanisms suggested by Persinger and Lafrenière to account for the lights was a form of piezo-electrical effect. This occurs when crystals are crushed under physical pressure, producing electricity and sometimes ionizing the air above to create glowing lights. It would be quite understandable for anyone witnessing these lights to interpret them as supernatural or extraterrestrial in origin.

The lights are assumed to be plasmas, charged gases that are often called the "fourth state" of matter. Plasmas exist in a number of forms—they help to make today's TV screens flatter and brighter, illuminate our cities through neon signs and light up the night as stars—in fact they make up the bulk of the known Universe. They are visible on Earth in auroras and, more rarely, as ball lightning. Airborne plasmas glow brightly at night and can appear silvery and metallic during the day—just like flying saucers.

One of the great mysteries surrounding these plasmas is that they have been described as behaving "intelligently"—responding to the actions of their human observers, or at least seeming to. This notion of intelligent gases has only really been the domain of the most far out science fiction, but research by Mircea Sanduloviciu and his colleagues at Romania's Cuza University suggests that life may, indeed, be a gas.

Sanduloviciu wanted to recreate the conditions on Earth before life began, in which the electrified atmosphere was highly conducive to plasma formation. He found that, when charged, the negative electrons in the gas atmosphere—in this case argon—formed a spherical outer layer of up to 3 cm (1.2 in) across, the positive ions created an inner layer and the remaining gas would form a nucleus at the center—and there you have an atom.

Such a distinctive outer boundary layer is one of the four attributes that define living cells. According to Sanduloviciu, the plasma spheres also display the other three: they replicate—by splitting in two—grow, and even "communicate" via electromagnetic energy. The plasmas could, he thinks, have been the first living matter on the planet. If Sanduloviciu is right, then perhaps today's earthlight witnesses are encountering their distant descendents. ⌐

050 INNER JOURNEYS

In 2005 Steve Currey of Provo, Utah announced that in the summer of 2006 he would lead a remarkable expedition. The 24-day trip was to begin at Murmansk, Russia. Participants would board a Russian Nuclear Icebreaker and head to the Geographic North Pole, from where they would travel 600 miles along Meridian 141 East to their final destination—the North Polar opening of the Hollow Earth. Sadly Steve died of a brain tumor in July of that year and the expedition was canceled. This hasn't stopped the other team members, who hope to complete the voyage under one Dr. Brooks Agnew sometime in the foreseeable future.

Their planned route is taken from a book called *The Smoky God*

The Cellular Cosmogony

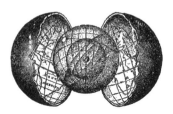

Discovered by Koresh
In 1870

(1908) by Wilis George Emmerson, which claims to be the truthful account of Norwegian fisherman Olaf Jansen's encounters with a race of ancient giants living inside the Earth. Most readers will recognize the book as science fiction—following earlier tales by Edgar Allen Poe and Jules Verne—yet for some Hollow Earth enthusiasts it remains startling fact.

Modern Hollow Earth ideas originated with US Army captain John Cleves Symmes. In 1818 he proposed that large entrances to the Earth's interior lay at both Poles. "I ask one-hundred brave companions to start from Siberia... with reindeers and sleighs, on the ice of the frozen sea; I engage we find a warm, rich land, stocked with thrifty vegetables and animals, if not men, on reaching one degree north of latitude 82," he wrote in the first of his many pamphlets on the subject. Two years later "Captain Adam Seaborn" printed *Symzonia: Voyage of Discovery*, claiming to have successfully followed Symmes' directions to the center of the Earth.

Symmes gained many adherents, the

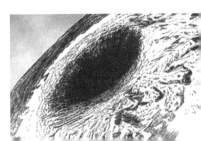

most enthusiastic being Jeremiah Reynolds, who almost persuaded US President John Quincy Adams to approve an official expedition, but the project was scrapped by his successor, Andrew Jackson. Reynolds did eventually lead a voyage of his own, sailing to Antarctica in October 1829—just months after Symmes' death—but they were unable to penetrate the ice surrounding the continent.

The late nineteenth century esoteric underground was well populated with Hollow Earth believers, most notably Cyrus Teed, whose theories would later influence Nazi cosmology. Unverified rumors of Nazi Hollow Earth expeditions persist to this day.

Our current geophysical outlook doesn't leave much room for an inhabited interior, but there's only one way to find out for sure: $19,000 buys a place on the Currey-Agnew expedition. See you at the pole!

051 DIG IT

While blasting rock at Dorchester, near Boston, Massachusetts, workmen found a zinc vase, inlaid with silver and intricately decorated with flowers and wreaths. According to *Scientific American* of June 5, 1852, the vessel was about 6 in (15 cm) wide by 4-1/3 in (11 cm) high. The rock from which it had apparently sprung is about 600 million years old—the tail end of the Precambrian period, whose worms and jellyfish aren't known for their delicate metalwork.

Meanwhile *The Morrisonville Times* of June 1891 (the exact date is not specified) relates that Mrs. S. W. Culp of Morrisonville, Illinois, found 10 in (25 cm) of eight-carat gold chain, of "quaint" design, firmly embedded in a large lump of coal. Her supply originated in a seam 260–320 million years old. There are many such reports, mostly in nineteenth and early twentieth century science journals and newspapers. Others document finds of iron nails, hammers, drill bits, screws, arrowheads, coins, pots, and bones in coal and ancient rock.

Michael Cremo and Richard Thompson, whose book *Forbidden Archaeology* collects a number of such tales, see the finds as proof that waves of human civilization have existed on the planet for hundreds of mil-

A Relic of a By-Gone Age.

A few days ago a powerful blast was made in the rock at Meeting House Hill, in Dorchester, a few rods south of Rev. Mr. Hall's meeting house. The blast threw out an immense mass of rock, some of the pieces weighing several tons and scattered small fragments in all directions. Among them was picked up a metallic vessel in two parts, rent asunder by the explosion. On putting the two parts together it formed a bell-shaped vessel, 4½ inches high, 6½ inches at the base, 2½ inches at the top, and about an eighth of an inch in thickness. The body of this vessel resembles zinc in color, or a composition metal, in which there is a considerable portion of silver. On the sides there are six figures of a flower, or bouquet, beautifully inlaid with pure silver, and around the lower part of the vessel a vine, or wreath, inlaid also with silver. The chasing, carving, and inlaying are exquisitely done by the art of some cunning workman. This curious and unknown vessel was blown out of the solid pudding stone, fifteen feet below the surface. It is now in the possession of Mr. John Kettell. Dr. J. V. C. Smith, who has recently travelled in the East, and examined hundreds of curious domestic utensils, and has drawings of them, has never seen anything resembling this. He has taken a drawing and accurate dimensions of it, to be submitted to the scientific. There is no doubt but that this curiosity was blown out of the rock, as above stated; but will Professor Agassiz, or some other scientific man please to tell us how it came there? The matter is worthy of investigation, as there is no deception in the case.

[The above is from the Boston Transcript and the wonder to us is, how the Transcript can suppose Prof. Agassiz qualified to tell how it got there any more than John Doyle, the blacksmith. This is not a question of zoology, botany, or geology, but one relating to an antique metal vessel perhaps made by Tubal Cain, the first inhabitant of Dorchester.

Scientific American, Vol. VII, number 38, p. 298
June 7, 1851

lions of years. This comes as less of a surprise when we learn that they are Hare Krishnas, for whom such great cycles of time are a matter of doctrinal belief. Christian fundamentalists, meanwhile, use the data to bolster Biblical accounts of creation. Other authors invoke extraterrestrials, time travelers, teleportation and similar futuristic fancies.

Sadly, those cases that can be investigated further often turn out to have less riveting explanations. Many objects turn out to be natural formations. Take, for example, the "2.8 billion-year-old" metal "cricket balls"—grooved spheres found in mines in South Africa's Western Transvaal during the 1980s. Considered by some to be the products of anomalous engineering technologies, they were identified by geologists as naturally occurring pyrite nodules, formed from clay or volcanic ash. This is not often mentioned in the mytho-archeological literature.

Other objects are exactly what they appear to be—it's their geological surroundings that are incorrectly identified. The "Coso Artefact," a metallic object found in a half-million-year-old geode near Olancha, California in February 1961, was described as looking like a spark plug. That's because it was a spark plug, manufactured by Champion in the 1920s. The surrounding material wasn't a geode after all, but a 40-year-old encrustation of clay. ⌐

A mong the wonders on display at London's 1862 Great Exhibition was a lump of coal dug from a seam 295 ft (90 m) below Newport in Monmouthshire, Wales. With it was a frog that miners claimed to have found alive, encased in a similar lump of coal, presumably millions of years old.

Their claim enraged the naturalist Frank Buckland, who demanded in *The Times* that the offending item be removed from display. As a result of the ensuing fracas, Professor Richard Owen, then Superintendent of the British Museum's natural history department, received so many specimens of toads and frogs found in rocks that he appointed his wife to deal with them.

Written records of animals, predominantly amphibians, found encased in solid rock date back to at least the sixteenth century. The usual story is that workmen digging in a quarry or mine find the creatures inhabiting a cavity roughly their own size and shape. Whether they fell down a crack which then sealed over, were dropped, flowed or blown there as frogspawn, as was once thought, or were even deliberately placed into the cracks by humans is anyone's guess. Of course in some cases, their discoverers may simply have made a leap of judgment on finding the creatures hopping around as they struck a particular stone or seam. More recent reports describe creatures living in concrete.

Several experimental investigations into the matter have been carried out. In 1771 the French naturalist Louis-Theodore Herissant entombed three toads in plaster cells that were themselves encased in wood. Two of the unlucky *Bufos* remained alive three years later. In 1825 the Oxford geologist, William Buckland found that several toads he had encased in limestone were still living a year later. Toad biology would support these two examples—the Sonoran Desert Toad, *Bufo alvarius*, for example, can spend many years hibernating in dry ground—though this hardly explains cases like that which so enraged Buckland.

Frogs, toads and other animals have also been found encased in trees. In his *Natural History of Selborne* (1789), the English naturalist Gilbert White describes a shrew found in this way. These were usually the handiwork of superstitious farmhands who placed them in tree holes as sacrifices, sealing them in with a peg of the same wood, which gradually grew over the hole. Such trees would then be considered healing sites.

Those still curious about this long-lived mystery will find a mummified toad, originally discovered inside a flint nodule in a Lewes quarry in about 1900, on display at the Booth Museum in Brighton, England. ⬭

053 ROCK SOUND

Were some ancient sites acoustically designed? Were they the first rock venues?

Ongoing research at Neolithic sites around the UK has revealed striking similarities in their acoustical properties. Key examples, both in Ireland, are the huge passage tomb of Newgrange and the burial mound known as Cairn L at Loughcrew. The two sites contain passageways leading to large circular chambers, and have a resonant frequency (at which sounds naturally echo and reverberate) of approximately 110 Hz—the frequency of the male baritone, the second lowest singing voice. Standing waves, whereby sounds are reflected off walls and superimposed onto one another, and other acoustic curiosities have been observed in these and several other sites. Stone circles including Avebury and Stonehenge also appear to reflect sound in distinctive ways.

Archeologists have suggested that chanting, singing and drumming at these sites would have produced reverberating echoes that may have been interpreted as voices of spirits or gods; they may also have induced physiological and psychological changes in people, adding to their potency as sites of spiritual importance. These acoustic discoveries might also shed light on some of the visual motifs etched into the walls of so many ancient sites. Experiments in a replica of the Newgrange passage at Princeton University showed that if a site was smoky or misty, standing sound waves would become visible as they vibrated particles in the air. Could this visualizing effect account for the numerous zig-zag and concentric ring markings on the chamber walls?

Intriguing acoustic effects have also been noted at sites in the Americas, from Anasazi *kivas* (ritual chambers) in New Mexico, to Chichen Itza on Mexico's Yucatan peninsula. Here, the famed Mayan pyramid of Kukulkan or Quetzelcoatl is best known for the way the solstices and equinoxes are reflected in its stones, but professional acoustician David Lubman has observed another aspect to its design. If you clap your hands in front of the pyramid, the sound is reflected back by its stone steps, sounding, Lubman claims, like the chirp of the quetzal bird, sacred to the Mayans.

Acoustic archeology is a young field that is finally gaining respectability within the academic establishment. New discoveries are being made constantly, so next time you find yourself at an ancient site, sing, clap your hands—and listen carefully. ▢

I n the mid-seventeenth century the famed architect Inigo Jones was instructed to survey the great ruins of Stonehenge in Wiltshire, England. He declared them far too sophisticated to have been built by the barbaric Druids, as was generally supposed, and proposed a Roman origin.

Almost a century later, the antiquarian William Stukeley overturned Jones' theory, noting the site's ancient age, and, more startling still, its astronomical alignments. Stukeley also suggested that the sites shared a uniform system of measurement. Whoever they were, our ancient monument builders were no barbarians.

Two hundred years later, such notions were largely forgotten, the stones a quaint picnic spot still untouched by hippy hands. But all that was about to change.

1967 saw the publication of *Megalithic Sites in Britain* by Alexander Thom, a retired Oxford engineering professor who had spent over a decade surveying megaliths throughout the land. Thom concluded that the megalith builders had used a standard, precise unit of measurement, which he called the Megalithic Yard (MY). This measured 2.72 ft (0.83 m, 40 Megalithic Inches, half a Megalithic Fathom), a number he reached through statistical analysis of 145 stone circles.

To Thom, the design, placing and likely function of the stone circles revealed that their builders had a profound understanding of topography, astronomy and the measurement of time. They demonstrated a sophisticated knowledge that would be lost for thousands of years, equaling that of medieval Europe.

Thom's book fit snugly alongside Gerald Hawkins' *Stonehenge Decoded* (1965)—which presented the monument as an ancient observatory —and John Michell's *The View Over Atlantis* (1967), which resurrected the largely forgotten theory of Ley Lines, originally developed by Englishman Alfred Watkins in the 1920s. Between them, the three authors unleashed a hoard of hairy dowsers and crystal-danglers onto the landscape, ensuring that our green and pleasant land would never be viewed in quite the same way again.

Predictably, Thom's theory has not convinced many mainstream archeologists. His critics point out inaccuracies of over one foot (30 cm) in some of his measurements of irregular and damaged circles, as well as fundamental problems with his statistical calculations.

But the Megalithic Yard has no shortage of supporters. Some trace its origins back to Atlantis; others see its length reflected in the motions of the heavens. But until someone uncovers the original megalithic ruler, Thom's theory is destined to remain a controversial one. ⬡

055 CAN YOU HEAR THE HUM?

I t's described as a low, modulated drone that sounds like an idling diesel engine or a distant aircraft. It starts and stops abruptly and is sometimes accompanied by headaches, nausea and other forms of physical discomfort.

It has been heard in Bristol, Birmingham, Hertfordshire and Strathclyde in the UK; in Taos, New Mexico and Kokomo, Indiana; as well as in Canada, Germany, Sweden, Denmark and New Zealand. Typically, fewer than 5 percent of people can hear it, with older women seemingly most likely to be affected. As well as innumerable sleepless nights, hums have been linked to at least three suicides in the UK.

The most frequently blamed sources are the Extremely Low Frequency (ELF) signals used to communicate with submarines, typically at around 76 Hz. ELF Transmitters are known to exist in Michigan and Wisconsin in the USA, and on Russia's Kola Peninsula. The first hum reports, however, predate ELF technology: in the 1940s, residents in London's Belgravia blamed a hum on a powerful radio transmitter in the Russian embassy's basement.

Other hums have been attributed to underground gas pipes, seismic activity, mating fish and underground military bases. Only two have been conclusively identified: in 2003, a low hum in Kokomo, Indiana was traced to industrial fans in a local factory, while in Auckland, New Zealand in 2006 a hum was recorded at 56 Hz, though its source remains unidentified.

The human audible frequency range is approximately 20 to 20,000 Hz. Women seem to be more sensitive to higher frequencies, becoming less so with age. Perhaps this makes older women more aware of lower frequencies, so fitting the hum demographic? That some hums have been measured by audio instruments would suggest that some sufferers are sensitive to frequencies usually unheard by the rest of us. One possibility is that the low and high frequency electromagnetic signals that constantly surround us might, in some cases, bypass the eardrums and affect the brain directly.

But cases of objectively measured hums appear to be the exception rather than the rule. When a German hum support group asked sufferers to identify the sound that they heard using a tone generator, they produced several different answers, indicating either a wide range of hums or, at least in some cases, an internal source—perhaps blood rushing past the ears. Stress certainly heightens people's sensitivity to the hum and also leads to high blood pressure, making it one possible cause for cases that aren't picked up on audio equipment, while increased local media reporting no doubt contributes to the spread of hum anxiety.

For now, the research and investigations continue while, for an unfortunate minority, the hum goes on. ⬤

056 THE FEAR FREQUENCY

Have you ever wondered what a ghost sounds like? The late English engineer Vic Tandy may have found out.

In the early 1980s, Tandy was working in a laboratory designing medical equipment. Word began to spread among the staff that the labs might be haunted, something Tandy put down to the constant wheeze of life-support machines operating in the building.

One evening he was working on his own in the lab when he began to feel distinctly uncomfortable, breaking into a cold sweat as the hairs on the back of his neck stood on end. He was convinced that he was being watched. Then, out of the corner of his eye, Tandy noticed an ominous gray shape drifting slowly into view, but when he turned around to face it, it was gone. Terrified, he went straight home.

The next day Tandy, a keen fencer, noticed that a foil blade clamped in a vice was vibrating up and down very fast. He found that the vibrations were caused by a standing sound wave that was bouncing between the end walls of the laboratory and reached a peak of intensity in the center of the room. He calculated that the frequency of the standing wave was about 19 Hz and soon discovered that it was produced by a newly installed extractor fan. When the fan was turned off, the sound wave disappeared.

The key here is frequency: 19 Hz is in the range known as infrasound, below the range of human hearing that begins at about 20 Hz. Tandy learned that low frequencies in this region can affect humans and animals in several ways, causing discomfort, dizziness, blurred vision (because your eyeballs are vibrating), hyperventilation and fear, possibly leading to panic attacks.

A more recent investigation took place in an allegedly haunted fourteenth century pub cellar in Coventry, Warwickshire, where people had reported terrifying experiences for many years, including seeing a spectral gray lady. Here Tandy also uncovered a 19 Hz standing wave, adding further evidential weight to his theory.

In 2006 Tandy's ideas were challenged by researchers Jason Braithwaite and Maurice Townsend, who have suggested that electromagnetic forces, possibly induced by infrasound vibrations, are more likely culprits than infrasound itself. However, in an interesting parallel, zoologists have recorded that, prior to an attack, a tiger's roar contains frequencies of about 18 Hz, which may serve to disorientate and paralyze their intended victims before they make a meal of them. ⌣

On February 19, 1994, 31-year-old Gloria Ramirez was brought into Riverside General Hospital in California suffering from chest pains and vomiting. As she drew a blood sample, nurse Susan Kane noted a "foul odor" before passing out cold, followed by Dr. Julie Gorchynski and four more staff, who collapsed moments later. Ramirez died shortly afterwards, while the affected staff were themselves hospitalized, some remaining chronically ill for months.

No entirely satisfactory toxicological explanation for the Ramirez affair has ever been put forward. To an eighteenth century scientist however, the answer would have been straightforward—the hospital staff had been struck down by "mephitic vapors" emanating from Ramirez's decaying body.

At this time, air was considered to be the binding substance that held the body together. Bad air—known as a miasma—was a sign of decay or sickness, and inhaling the miasma of a corpse could lead to illnesses including syphilis, scurvy and gangrene.

Identifying miasmas became something of a scientific obsession, leading to the sudden importance of that finely tuned scientific instrument, the nose. In 1766, Madame Thiroux d'Arconville published *Essai Pour Servir à l'Histoire de la Putréfaction,* a magnificently detailed, 600-page account of the putrefaction of 300 substances, according to the season and weather.

Miasmas were also present in the very earth, emerging naturally due to high temperatures underground. Mines and quarries were considered inherently unsafe, with marble and other stones giving off potentially dangerous vapors, as were fields—freshly tilled soil was deemed especially dangerous. Earthquakes, like that which struck Lisbon in 1755, presented scenes of miasmic catastrophe. Ordinary walls were also thought capable of absorbing sickness, with prisons and hospitals being worst affected, some being shut down to prevent further malady.

Concerns about pollution, if biologically misguided, shared similarities with our own today: some areas were considered so afflicted by human and animal waste that future generations would be unable to live in them. As London and Paris grew more densely populous, the problem of waste became ever more pressing. Things came to a putrid head for eight weeks during June and July of 1858 with The Great Stink of London, when Parliament was abandoned for a short time; the stench from the River Thames was so awful that its members couldn't bear to stay in the building any longer.

In London, the morbific miasmas dissipated following the construction of 82 miles of sewers and the Embankment along the Thames, while in France, Louis Pasteur's germ theory revealed the mephitic odors to be one of several possible symptoms of sickness, rather than a cause.

But we still don't know what happened to Gloria Ramirez. ◌

058 THE FOG

"The summer of the year 1783 was an amazing and portentous one and full of horrible phaenomena," writes the Hampshire naturalist Gilbert White in *The Natural History of Selborne* (1789).

For about a month between June 23rd and July 20th, English skies were thick with a "peculiar haze" or "smoky fog," while terrible lightning storms left people cowering in their homes. One storm provided five men with a stay of execution at Tyburn as the gallows, and the assembled crowd, were flooded.

Noted White: "The Sun, at noon, looked as blank as a clouded Moon, and shed a rust coloured ferruginous light on the ground, and floors of rooms; but was particularly lurid and blood-coloured at rising and setting."

Despite winds that seemed to change direction with alarming frequency, the country was engulfed in a heat so stifling that meat was said to rot within a day and the air was filled with clouds of flies.

It was a turbulent year for planet Earth. In February, cataclysmic earthquakes and volcanic eruptions killed 30,000 in Calabria, Italy, while between May and August Japan's Asama-yama volcano disgorged enormous quantities of fire and brimstone, killing about 1200 people.

The Great Fog of July resulted from one of the hugest volcanic eruptions in recorded history, when Iceland's Skaptar Jokull erupted on June 10th, creating a lava flow so large that it turned the entire Skapta river to steam and filled a nearby lake. Further eruptions caused flows 50 miles (80 km) long and ten miles (16 km) wide. 9500 people died, one fifth of Iceland's population. Most of them were asphyxiated by a colossal dust cloud that eventually drifted south, covering all of Europe from Scotland to North Africa.

As the cloud settled over England, many began to fear that the end was finally nigh, with apocalyptic fervor climaxing on August 18th when a huge meteorite—then still something of mystery to science —was logged by observers from Sheffield to Canterbury before exploding spectacularly over the English Channel.

Gilbert White wasn't the only person watching the skies that summer. An 11-year-old Oxfordshire boy named Luke Howard was inspired by the extraordinary cloud to start studying ordinary ones, and as an adult he would give them the names by which they are still known today. ▢

T he recent discovery of polar ice sheets on the Moon reminds us that our satellite Selene still has her secrets; but another of her mysteries has been visible from the Earth's surface for almost 500 years.

In 1540, over half a century before the first telescope was unveiled, stargazers in the German city of Worms saw a star-like object appear in the Moon's northeastern Calippus region. 110 years later, the Polish astronomer Hevelius noted the appearance of a Mons Porphyrites, or "red hill," in the northwestern crater Aristarchus. In April 1787 the astronomer Sir William Herschel saw so many red lights in this same crater that he thought he was witnessing a lunar volcano and summoned King George III to join him in observing them.

Since those early days, NASA records reveal that several hundred spots and flashes of light,

colored glows, mists and unusual shapes and shadows have been observed on the lunar surface. Collectively they're known as Transient Lunar Phenomena (TLP), a term coined by keen TLP observer, the famed British astronomer Sir Patrick Moore.

Professional astronomers sniffed at these reports for many years, declaring them optical illusions or telescope errors. But a 1968 NASA catalog documented hundreds of sightings, the vast majority in and around Aristarchus. In July the following year, as Apollo 11 entered lunar orbit, earthbound German astronomers reported seeing a bright TLP in Aristarchus. Asked to check it out, Neil Armstrong noted unusual illumination in what he took to be the same crater.

So is the Moon, supposedly a lifeless chunk of basaltic rock, undergoing periodic volcanic activity? Most astronomers don't think so. In 1963 Zdenek Kopal observed that a significant TLP event occurred during a major solar flare-up. He suggested that particles from these vast electromagnetic emissions caused moon rocks to glow, particularly when the Moon is passing through the Earth's own magnetosphere.

Another theory was put forward by Audouin Dollfus of the Paris Observatory in 1992. For several days from December 30th he watched a number of glowing clouds emerge from Langrenus, not usually considered a TLP hotspot. Dollfus later identified them as moon dust, thrown up into the sunlight by underground gaseous emissions. The crater's surface is indeed cracked, and Radon gas was detected in the lunar atmosphere during the Apollo missions. Perhaps these moon burps are the source of some TLP, but it'll take at least another lunar landing to find out for sure, and who knows when that's going to happen? 🗗

THE ARCHITECTS OF MARS

No sooner had NASA published images of an ice lake inside a crater on the Martian plain Vastitas Borealis in July 2005, than Internet exo-archeologists were excitedly pointing out the crumbled ruins of a vast, ancient city on the crater's banks.

Earthlings have been seeing things on Mars since at least 1877, when the Italian astronomer Giovanni Schiaparelli noted long, straight lines, which he called *canali*, on the planet's surface. Translated into English as "canals," the markings led respected astronomer Percival Lowell to propose the existence of a network of Martian irrigation canals, indicative of an advanced civilization not unlike our own.

99 years later the Viking 1 Space Probe beamed back images of Cydonia, what may have been a coastal region in Mars' northern hemisphere. Amongst several mountainous protrusions was what appeared to be a blank, humanoid face wearing a hint of a smile and a crash helmet.

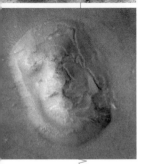

The image was singled out by Vince DiPietro and Greg Molenaar, who cleaned it up and represented it to the world a few years later, this time with the addition of a large pyramid. Their find drew the attention of sometime NASA consultant Richard Hoagland, who identified more anomalies in the region, including walls, a "fort," several smaller pyramids and something resembling the 4000-year-old, man made mound, Silbury Hill in Wiltshire, England. Through some fantastical over-interpretation by zealous mytho-archeologists, the Cydonian structures were soon being connected to human sacred sites at Avebury, the stone circle nearest to Silbury Hill, and the Pyramids and Sphinx at Giza, in Egypt.

When NASA's Mars Global Surveyor returned to the red planet in 1996, they agreed to settle the debate by re-photographing the "face," once in 1998 and again in 2001, producing images ten times sharper than Viking's originals. To the dismay of its admirers, the images showed that the face was actually a raised plateau. While some die-hards protested that NASA's secret space fleet had nuked the real face to prevent its exposure, most agreed that the Cyberman of Cydonia was a natural formation brought to life by a trick of the light and 1970s digital imaging technology.

But, as NASA continues to explore Mars, so do the armchair exo-explorers, Arthur C. Clarke amongst them, who have now added colossal sand worms, beetles, numerous fossils and swathes of vegetation to the increasingly crowded Martian environment.

President George W. Bush has announced that NASA intends to send humans to Mars by 2030—let's hope they pack a camera. ◻

061 BY THE LIGHT OF THE SILVERY MOON

"Avoid the thirteenth of the waxing month for beginning to sow: yet it is the best day for setting plants," advises the eighth century B.C.E. Greek poet Hesiod, in his paean to good honest labor, *Works and Days*.

Hesiod's is the first written reference to planting crops according to lunar—or synodic—cycles, as was practiced throughout the ancient world. For centuries, planting crops by the phases of the Moon was an integral part of gardening folklore, and it remains a secret weapon for an increasing number of organic growers, horticulturists and wine makers. Science may still be skeptical of such green-fingered lunacy but many respected gardeners, like John Harris at Cornwall's Tresilian House Gardens, swear and sow by it.

The twentieth century's resurgence of interest in the technique was evolved from biodynamic agricultural techniques promoted by the Austrian mystic Rudolph Steiner, who encouraged farmers to consider their soil and crops as part of a larger, living system. During the 1920s and '30s Steiner's colleague Lili Kolisko conducted trials to demonstrate the value of Steiner's ideas. She found that beans, cabbage, cucumbers, lettuce, peas and tomatoes sown two days before the full moon geminated more successfully, grew more vigorously and generated higher yields than those planted before a new moon.

Following up Kolisko's work in the 1950s, Maria Thun sewed radishes every day of the growing season and compared their output. Thun launched Kolisko's ideas into the realm of astrology, claiming that different types of plants benefited from being sewn when the Moon was in specific astrological "houses." In the 1970s, market gardener Reg Muntz and astrologer Colin Bishop reported increased yields of up to 50 percent by following lunar cycles.

They may sound a little wacky, but there is a scientific basis for these claims. Just as the Moon affects tides, it also affects moisture in the soil, which is drawn towards the surface during the full moon. Agronomist Nick Kollerstrom found that seeds planted at the full moon germinate more quickly—perhaps because of this increased moisture—although he remains skeptical about claims of increased yields.

In the 1940s, Yale biologist H. S. Burr found that moon phases affected the electrical potential of trees, which he attributed to changes in the flow of sap. More recent research has demonstrated that trees swell and shrink according to the Moon's position, even in total darkness. If there's a moon tide in trees, why can't other plants also be affected? 🗗

HOT ROCKS

hese days we all know that meteorites originate in space, caused by cosmic debris—often as small as a grain of sand—entering the Earth's atmosphere. But until two hundred years ago the scientific establishment considered this to be an outrageous notion despite the small mountain of evidence that was literally falling at their feet.

Even in the late-eighteenth century the prevailing belief amongst the scientific elite was that space was empty and rocks did not come from the sky; to declare otherwise was superstition or madness. Even in 1768, when a scientist found a still-smoking rock on the ground, it was decreed that it must have been struck by lightning. The realization that these stones actually came from outside our atmosphere took decades to sink in and was nothing short of a scientific revolution.

As late as 1771, the statements of scientists from Paris to Sussex who observed a meteor hurtling across the English Channel were not enough to sway orthodox opinion. But the mood was slowly changing. By 1794 the German physicist and astronomer Ernst Chladni had amassed evidence to show that meteorites did indeed come from space. Adding weight to his testimony, the British chemist Edward Howard noted in 1802 that these "aerolites" shared similarly unusual compositions, including the presence of nickel, which had first been extracted in 1751. The final piece in the puzzle came in 1803 when French scientist Jean-Baptiste Biot saw stones falling from a fireball over Normandy. Faced by such a wealth of fresh evidence from different disciplines, the prevailing orthodoxy was finally, albeit reluctantly, swayed.

One irony of all this is that ancient cultures were undoubtedly familiar with meteorites. By around 4000 B.C.E., the Egyptians and Sumerians, who both associated the rocks with the heavens, were extracting iron from them for use in ritual weapons and objects. Meteorite worship was practiced for centuries all over the world—famously, for example, at the temples of Apollo at Delphi and Diana at Ephesus—and still is. Many believe that the Hadschar al Aswad, the black stone in the Ka'ba wall at Mecca, which had been venerated for centuries before the time of Mohammed, and is still worshipped by tens of thousands of pilgrims every year during the Hajj, is actually a meteorite.

The ancients may not have fully understood the stones, but at least they knew where they were coming from. ▢

"Of several Accidents that were reported to have attended its Passage, many were the Effect of Fancy, such as the hearing it hiss as it went along, as if it had been very near at hand…"

Edmond Halley, later England's Astronomer Royal, is here describing eyewitness accounts of a large meteorite—then still a puzzling phenomenon—seen on March 19, 1719. A few years earlier Halley had analyzed several simultaneous sighting reports, estimating those meteors to be 60 miles up, hence his dismissal of the alleged audible effects.

In fact people have been hearing meteorites for centuries. A Chinese text from 817 C.E. mentions a sound like a "flock of cranes in flight" (presumably their whooshing wings, rather than their vocalizations), while an Arabic record of 1026 C.E. details a loud sound accompanying a particularly bright meteor. Similar observations were made throughout the eighteenth and nineteenth centuries, but the more scientists understood about meteors, the more likely they were to dismiss such phenomena as hallucinations.

The Russian mineralogist Peter Dravert first discussed "electrophonic bolides" in 1940, borrowing a term used to describe the sound of an electrical current passing through the head; while A. G. Kalashnikov measured 1 Hz ULF emissions from a bolide in 1949. However, a 1963 RAND Corporation catalog, *Anomalous sounds and electromagnetic effects associated with fireball entry,* was unable to suggest a physical mechanism for the sounds, so its authors stuck to a psychological explanation.

But the evidence is mounting. Physicist Colin Keay interviewed numerous witnesses who heard a huge meteorite as it lit up the night skies of New South Wales, Australia in 1978. They described sounds like hissing steam, popping fireworks and electrical crackling as the fireball rushed overhead for 20 seconds, brighter than the full moon. Keay suggested that meteoric VLF emissions, traveling at the speed of light, are converted into sound by a transducer near the witness: this can be anything from an antenna to somebody's spectacles or even frizzy hair. In 1998, a Croatian team stationed in Mongolia finally captured a bolide's "pop" on video, complete with soundtrack.

The mechanism behind these sounds remains a mystery. The

Croatians think the answer lies in the interaction between the meteor and the Earth's ionosphere, while Keay sees it in the bolide's twisted plasma trail. Meanwhile, as electrophonic phenomena gain acceptability, the range of possible sources identified with these enigmatic natural sounds has expanded to include auroras, lightning, earthquakes (perhaps explaining animal "predictions"), mystery "hums" and close-passing comets. 🗆

064 SPACE GUNK

" I t appeared larger than the Sun, illumined the hemisphere nearly as light as day. [And when it fell]… a large company of the citizens immediately repaired to the spot and found a body of fetid jelly, four feet in diameter." *Scientific American,* 1846.

This description of a spectacular meteorite fall is a fine example of the phenomenon named by Welsh shepherds *pwdre* (sometimes *powdre*) *ser* —the rot of the stars—also known as star slough, star shot, star spawn or star jelly. These gelatinous blobs, usually whitish, translucent and foul smelling, have been associated with meteorite falls for centuries. In 1656 the philosopher Henry More, observed that "the Starres eat… those falling Starres, as some call them, which are found on the earth in the form of a trembling gelly, are their excrement."

Often found in early mornings, the jellies usually dry up quickly, disappearing to almost nothing as the day warms up.

Most meteors, composed of rock and ore, burn up instantly on hitting the Earth's atmosphere, so gelatinous material wouldn't stand a chance—this is certainly not space gunk. Since at least the early-eighteenth century, the most common earthbound explanation for the mystery goo has been that it is something vomited up by birds or animals; the Welsh naturalist Thomas Pennant, writing later that century, considered this the answer.

Currently popular is the idea that the gray gloop is frog spawn barfed up by amphibian-eating creatures, though no frogs' eggs have ever actually been identified within it, and most finds are a good deal larger than your average frog. A recent refinement of the concept is that if a frog is swallowed prior to ovulation, its regurgitated egg duct—which swells dramatically when wet—holds the properties necessary to identify it as *pwdre ser*. It all seems very complicated.

But that doesn't mean that jellies never fall from *within* the atmosphere, as frogs, fish and other critters are occasionally wont to do. In 1995 a translucent jelly like substance—"enough to fill a kettle" according to its lucky finder—was discovered in a garden in Horley, Oxfordshire. In 1983, Reading Massachusetts was pelted with a grayish-white jelly which, when analyzed, proved not to be waste from an airplane, as was at first assumed. Clearly better lab analysis is what's required, though we'd prefer that you didn't send your mystery goo samples to us.

065 STRANGE RAINS

On August 6, 2000, it rained fish over the Norfolk seaside town of Great Yarmouth in the eastern UK. Thousands of freshly dead baby sprats, about 2-inches (5 cm) long, littered roofs and gardens following the downpour. Meteorologists made the sensible assumption that the fish had been dumped there by a waterspout, although no one reported seeing one.

Falls of living creatures have been recorded for centuries. Pliny the Elder, in his *Natural History* of 77 C.E., cataloged those involving a number of animals and suggested that the seeds of the creatures are contained in dust or slime carried by the winds. The most common skydivers are fishes and frogs, but insects, worms, jellyfish, crabs and starfish have also experienced the joys of freefall. Other reports, while authentic, stretch the bounds of incredulity featuring plummeting lemmings, rats, even alligators—one being found, following a loud splash, in the ballast bag of an airborne dirigible during World War II.

Tornadoes and waterspouts remain the most likely culprits for the

majority of falls, but several questions remain. Some falls are intensely localized, appearing as if the creatures were dumped onto only one spot of ground, while often the critters involved are still alive, as if gently placed there, rather than dropped from the heavens. Falls almost always involve only one species; this can, in part, be explained: the wind picks up a load of material, then sifts it, leaving behind only those of a certain weight or shape. But why aren't other creatures or objects of similar proportions also found at fall sites? And, as in the Yarmouth case, there are few reports of witnesses to such freakish whirlwind phenomena. As Charles Fort, who was fascinated by these occurrences, observed: "A pond going up would be quite as interesting as frogs going down."

As well as living matter, inorganic substances also crash to earth, including seeds, stones and sand. However, these seem rather puny in comparison to the mighty ice blocks, several of which, weighing up to 9 lb (4 kg), fell on Spain during January 2000. And even they are dwarfed by the 440 lb (200 kg) titans that struck Brazil in 1995. The forces behind these monsters, involving sudden drops in atmospheric temperature caused by high concentrations of ions and aerosols, are only just beginning to be understood. In the meantime, as ever, watch the skies. And mind your head.

A s several international probes struggle valiantly to reach the relatively neighborly planet Mars, a small coterie of astrophysicists are quietly considering how humankind might venture beyond our own solar system.

The key to the problem is power: what kind of fuel will be both stable and plentiful enough to take a crew of humans into deep space? Among the more sober possibilities are ion propulsion, using xenon gas as a fuel, which sent NASA's Deep Space 1 probe a respectable 185 million miles or so, and solar sails, blown by photons from the Sun. But neither of these is ideal; xenon, though stable, is exhaustible, and solar sails would only be able to carry very small, light craft.

One theoretical energy source, however, would fit the bill perfectly. It's as accessible in outer space as it would be in the Outer Hebrides, because it exists, according to its advocates, everywhere, immersing everyone and everything in a foaming sea of energy.

Early quantum physicists theorized that all space, even the vacuum of outer space, contains a constantly bubbling field of electromagnetic energy, quantum fluctuations thought to be created by "virtual" photons constantly winking in and out of existence. This is Zero Point Energy (ZPE), so called because it would still exist at absolute zero—minus 273 ℃

$$E = \frac{\hbar\omega}{2}$$

(minus 459 °F)—when the atomic motions that generate thermal energy are at their slowest. Physicists John Wheeler and Richard Feynman calculated that there is enough such energy in the vacuum inside a single light bulb to boil all the world's oceans. The challenge, currently being undertaken by several teams, is how to tap it.

Some researchers have also suggested an intriguing connection between ZPE, inertia and gravitational pull. The push you feel while slowing down or turning when driving may actually be caused by ZPE fluctuations. A greater understanding of how to manipulate ZPE could one day lead to the control of gravitational and inertial forces, leading to new forms of propulsion and a revolution in space travel.

The number of physicists studying ZPE is small, and most are still operating at the theoretical level, but a breakthrough could one day provide the Ur-energy of the future. 🗆

067 KERANOGRAPHY

Wells Cathedral, Somerset, 1596: A lightning bolt strikes the building during a service, flinging members of the congregation to the ground. Recovering their senses, they make a shocking discovery. Noted a contemporary historian: "The marks of a cross were found to be imprinted on the bodies of those then at divine service... the Bishop himself found the marks upon him, and others were signed on the shoulder, the breast, the back and other parts."

Morgantown, Kentucky, 1872: A woman is standing in front of her window when she receives a shock from a lightning strike. Soon after she finds an image of the tree outside her window imprinted onto her breast.

Jonesville, Michigan, 1887: Farmer Amos J. Briggs is shooing cats away from his woodpile when it is struck by lightning. The cats die instantly. Briggs' watch explodes and his clothes are shredded. Returning inside, his wife is horrified to find the silhouette of a startled cat imprinted onto his bald pate. The image fades after two days.

Until the late nineteenth century, it was believed that the intense flash of heat and light produced during a lightning strike could burn the silhouette of nearby objects onto the skin of people in its vicinity. The phenomenon was taken seriously enough by scientists to be given the name keranography, though it's not a word you will hear often today.

Lightning travels at 186,000 miles per second with a current of between 10,000 and 40,000 amps. It hits the earth about 75 times a second, killing hundreds of people a year. Although only about 10 percent of strikes are fatal, the majority result in physical or psychological damage.

A typical lightning bolt has a diameter of only about 1.5 in (4 cm), but hits so fast that most strikes on humans don't leave visible wounds. The real damage is often internal, to the brain, heart or nervous system, which can be cooked and ready to eat in a split second. Those burns that do occur tend to be caused by metal objects close to the skin, such as jewelry or coins, while clothing and other items on the body can also be damaged.

When lightning burns are actually visible, they tend to be linear and branch-like, and it seems to have been these that lead to the accounts of images of trees, crosses, even startled cats, being captured on the body. ⌑

068 OCEANIC AURORAS

It's often said that we know more about outer space than we do about the depths of our oceans. Certainly they are home to a number of little-understood natural phenomena, of which none can be more dazzling than the lightwheels.

The Marine Observer, the journal of the Marine Division of the British Meteorological Office, contains detailed reports of lightwheel observations going back more than a century. Despite this, no expeditions have set out specifically to study them.

These spectacular submarine lightshows range in size from a few feet across to filling the entire visible ocean with phosphorescent colors. They take many forms: long, straight, undulating bands that stretch from horizon to horizon; simple rings and ellipses; concentric circles that pulsate from a brightly glowing center, as if transmitting some kind of signal; curving, spoked wheels whose arms rotate clockwise or anticlockwise, sometimes at dizzying speeds. The displays can last a few minutes or several hours, with numerous patterns appearing in sequence, or even simultaneously.

Ordinary bioluminescent phenomena occur most often in warmer waters like the Persian Gulf, the Indian Ocean and the South China Sea. Generally occurring between March and June, they consist of millions of microorganisms—single-celled bacteria and dinoflagellates—and larger plankton, which drift near the ocean surface. These microorganisms glow when shaken or disturbed, lighting up the wakes and bow waves of ships, or the crests of waves, and sometimes forming great milky white blankets over the sea. More uncommon are bright balls of light and long worm-like shapes seen glowing underwater—

as described by Thor Heyerdahl during his 1947 Kon-Tiki voyage—which have been attributed to shoals or large individual fish moving amongst the luminous organisms.

But the huge rings and wheels appear relatively rarely, and their cause remains undetermined. Some have connected the pulsating signal-like patterns to the communications devices on ships and submarines, but historical lightwheel reports predate radar and sonar, even radio. Witnesses have often tried turning their electrical equipment and engines on and off, with no specific effect on the lights. Others have attributed the patterns to whale communication, tectonic movement, geomagnetic pulses from the earth's core or, more fancifully, to the beacons of some great submarine, possibly alien, civilization.

Awe-inspiring and beautiful, unfilmed and barely photographed, the lightwheels remain an unfathomable mystery. ▢

PART FOUR

THE HUMAN CONDITION

"There's a young student at this university," neurology Professor John Lorber of Sheffield University told *Science* magazine in December 1980, "who has an IQ of 126, has gained a first-class honors degree in mathematics, and is socially completely normal. And yet the boy has virtually no brain."

A scan revealed that the student had only 1 mm of brain tissue lining the inside of his skull—where the rest of his brain should have been was filled with fluid. His was an extreme case of hydrocephalus—"water on the brain"—whereby cerebrospinal fluid fills the brain instead of circulating around it. Most sufferers can lead normal lives if regularly treated.

But if he had no brain, where was his mind?

Similar questions are raised by cases of "transplant memories." In 1988, Claire Sylvia received a heart and double-lung transplant. Following the operation she underwent some apparent personality changes: she began to have unusual (for her) cravings for beer, green peppers and chicken nuggets; she dreamt about beautiful women and experienced homosexual urges. She also dreamt of meetings with a young man called Tim. Alarmed,

Sylvia sought out her donor's family and discovered that her new organs had belonged to an 18-year-old boy. His name was Tim. Tim had a penchant for the same foods she was craving—he was actually eating chicken nuggets when he died—and Sylvia felt he was the boy in her dreams.

In the nineteenth century, the German anatomist Leopold Auerbach observed a complex network of nerve cells in the human digestive tract. This nerve bundle, a "second brain" containing more nerve cells than the spinal cord, was recently rediscovered by Michael Gershorn at Columbia University. Professor Wolfgang Prinz in Munich has also studied this second brain, and thinks that it could govern some of our emotional and physical responses to thoughts and events—hence, perhaps, "gut feelings."

Georgetown University's Dr. Candace Pert has suggested that neuropeptides are linked to our sense of self. These chemicals, found in all our major organs and muscles, enable communication between the mind and body. Pert's theory is that they also carry our emotions, and our memories. Is consciousness diffused through the body with them? ▢

Today, communication with our Martian probes is all about radio and can be a touch-and-go affair—witness several lost probes over the decades. Perhaps frustrated NASA engineers might like to hark back to a time when receiving messages from Mars was as simple as putting pen to paper.

In the late nineteenth century, at least four mediums were in regular contact with the Red Planet, the most celebrated being Catherine Elise Muller, known as "Helene Smith."

In 1890s Geneva, Muller conducted private séances in which she would talk, through her spirit guide Leopold, to the likes of Victor Hugo—in life also a keen spiritualist—and the Sicilian occultist Count Cagliostro. She also recalled past lives as Marie Antoinette and an Indian princess.

Muller's Martian breakthrough came after her spiritualist circle had been discussing astronomer Camille Flammarion's popular book *La planète Mars et ses conditions d'habitabilité* (1892). Via Leopold, who now lived there with several other spirits, Muller began to produce reams of information about Mars life. She claimed that, during her trances, her astral body was actually transported to the planet, so she was able to draw detailed Martian landscapes and to speak and write in its language. Through these visits, Muller built up a great body of material about day-to-day Mars life, full of fascinating details such as that gardening was a popular pastime, and that people traveled using lantern-like jet packs.

Word of Muller's Martian contacts spread and her case drew the attention of the psychologist Theodore Flournoy. He spent five years studying Muller, researching her family history and vision-prone childhood as well as attending séances and psychoanalyzing her. Although impressed by Muller as a person, Flournoy regarded her experiences as a marvel of psychology, rather than spiritualism. One of his key findings was that, while Muller's Martian had a consistent 23-letter alphabet, grammar and syntax, it was in fact a twisted variant of French, around which she had developed her new language.

Flournoy published the results of his investigation in *From India to the Planet Mars* (1899), protecting Muller's identity by renaming her "Helene Smith." Despite his skeptical deconstruction of her story, he expressed an admiration for Muller's prodigious feats of imagination and described her story as a case of idealism against grey reality," he wrote. "On the wings of dream, the individual flies, hoping to escape the thousand and one discouragements of the prosaic everyday." Mars here we come! 🖰

I n a 2001 study, over 60 out of 344 Dutch patients who were declared clinically dead and then resuscitated could recall aspects of their near-terminal experiences. Several reported elements of the classic Near Death Experience (NDE), including tunnels of light, floating outside the body and seeing their life flash before their eyes. All the subjects were interviewed within a week of their experiences and remembered clearly what had happened.

Variously interpreted as glimpses of the afterlife, or the soul's journey from the body, these experiences are helping researchers shed new light on the ultimate darkness.

Psychologist Dr. Susan Blackmore suggests that the "tunnel of light" described by so many NDE survivors results from a final surge of activity in the oxygen-starved brain's visual cortex. A computer simulation showed that when a visual signal is gradually overloaded, most cells trigger in the center of the visual field, where we see clearest, and fewer on the outer edges, where our vision is hazy. This creates the illusion of moving through a tunnel, which is ultimately filled with white "light."

Meanwhile, psychology professor Drouwe Draaisma has compared historical and contemporary reports of people's lives flashing before their eyes. He found accounts altered significantly over time, reflecting their subjects' beliefs, culture and technology. So eighteenth or nineteenth century experiences are usually described as a series of still images, while modern ones more resemble sequences from films. Moral elements, such as a judgment of one's life, seem to be a matter of personal taste.

Most puzzling are the Out of Body Experiences (OBE), whereby clinically dead patients describe floating above their bodies, sometimes accurately recounting events witnessed from this topsy-turvy perspective. Swiss researchers claim to have pinpointed how OBEs are created in the brain. Neurologist Olaf Blanke found that electrically stimulating the angular gyrus in the brain's parietal lobe—which maps your body's relationship to its surroundings—creates a dissociative state in which subjects feel that they are looking down on themselves. The sensation can also be induced by stress or a lack of blood in the brain. Perhaps this, combined with some outwardly indiscernible sensorial activity can build a mental image of the subject's surroundings during brain death?

While no amount of science can demystify death, as resuscitation techniques continue to improve, more of us are likely to discover how it feels to die and be able to report back from the other side.

When the Vatican canonized Padre Pio Forgione in 2002, they studiously avoided mentioning his most famous physical features: the iconic stigmata—or "wounds of Christ"—that appeared on his hands and left side.

St. Francis of Assisi, whose wounds appeared during an angelic vision in 1224, was the first recorded stigmatic. His unusual markings incorporated both Christ's wounds and the nails that inflicted them: less pious observers today might describe these as large scabs.

There have been about 400 recorded stigmatics since then, and about 25 remain bleeding today. The majority are women and virtually all of them are Catholic. Wounds appear most commonly on the hands and feet, but also on the sides of the body—where Jesus was speared while on the cross—and on the forehead, representing the crown of thorns. Most stigmatics bleed little, though some squeeze out up to a pint at a time.

The personality profile of the typical stigmatic is not a happy one. Many contemporary subjects have been victims of abuse and suffer from extremely low self-esteem. Other psychological problems are common, including eating disorders, which can themselves lead to severe bruising and hemorrhaging. Most stigmata—as with suspected frauds like St. Pio and the Bavarian mystic Theresa Neumann—are deliberately self-inflicted, or follow more complex patterns of self-harm, like those of Munchausen's Syndrome.

One common criticism of stigmata sufferers is that their wounds correspond more closely to religious art than to historically accurate Roman crucifixion techniques: Christ would have been pierced through the wrists and ankles, rather than his hands and feet. Some new wave stigmatics, like 29-year-old Argentinean Emiliano Aden, counter this by displaying correctly placed wounds.

Not all stigmatics are necessarily frauds; there are medical conditions that might account for some of them. Haemathidrosis, whereby a person appears to sweat blood, can occur at times of extreme stress; while spontaneous hemorrhages, called *psychogenic purpura* have also been recorded. Physical wounds have also manifested under hypnosis. In 1933, Dr. Alfred Lechler recreated the full range of stigmatic markings on a 29-year-old German peasant, after she had seen a film of Christ's crucifixion. Hypnosis has also been shown to help hemophiliacs with their condition, while other subjects have learned to direct the flow of blood to specific parts of the body. The key to stigmata is to be found not with God then, but deep within man. ◻

"The possibility of scientific annihilation of personal identity, or even worse, its purposeful control, has sometimes been considered a future threat more awful than atomic holocaust... These objections, however, are debatable."

So wrote Dr. Jose Delgado in his 1969 book *Physical Control of the Mind: Toward a Psychocivilized Society*. Delgado documents the myriad potential applications of Electrical Stimulation of the Brain (ESB), from helping the blind to see again to keeping criminals and dissidents under remote control.

The Spanish neurologist's hopes rested on a device he called the "stimoceiver." Once inserted into the appropriate part of the brain, the remotely operated stimoceiver could stimulate it with tiny electric pulses. In a dramatic demonstration in the early 1960s, Delgado entered a bullring and, at the press of a button, stopped a charging bull dead in its tracks. Delgado saw great potential in his creation, but he did note one potential problem: "the existence of wires leading from the brain to the stimoceiver outside of the scalp... could be a hindrance to hair grooming."

Modern day proponents of the mind control conspiracy use Delgado's stimoceivers to support their suspicions; but how far has the technology advanced?

In 2004 American scientists demonstrated a team of remote controlled rats that could be controlled from a laptop. Electrodes in the rats' brains activate their pleasure centers while steering them left or right. A tiny backpack acts as a receiver. Sanjiv K. Talwar, the team's leader, was

happy to concede that "the idea sounds a little creepy," but pointed out that the rats could save many human lives, for example in mine clearance.

The robo-rats were presented to the media as the very latest in biotechnology, but in his memoirs the late neurologist John Lilly—who began his career worked on the electrical stimulation of animal brains—recalled seeing a military film of a donkey being remotely steered up a hillside in the 1950s, using technology that he had helped to develop. The military plan was for these robot donkeys to carry large explosive devices, in effect making them remote control suicide bombers.

Whether or not we have the technology to control another human mind, brain implants are now being used to help the victims of paralysis. Dr. Philip Kennedy of Neuro Signals has developed an implanted device that allows JR, a paralyzed 53-year-old volunteer to move a mouse cursor around on a computer screen using thought alone. There are no wires—a small antenna, connected to the implant, pokes out of the top of JR's skull.

We may be approaching Delgado's psychocivilized society, but that hair is going to be a problem for some time yet. 🗪

074 PEER PRESSURE

What would it take for you to distrust the evidence of your own eyes? Only seven other people, according to a 1950s study conducted by the psychologist Solomon Asch.

Interested in the extent to which the pressure to conform affects our judgment, Asch devised a simple but devastatingly effective experiment. The test subject sits in a room with seven other people. The experimenter shows them an image of a vertical line, X, followed by three more lines, A, B and C, one of which is the same length as X. The people in the room are asked, one at a time, to state which of lines A, B and C is the same length as line X. The process is repeated several times during the session.

Initially, everybody in the room selects the correct line, but over the course of several rounds, the others begin to choose lines that are quite clearly not the same length as line X. In fact, the other seven people in the room are in cahoots with the experimenter. Six of them are always asked to make their choices first, giving the test subject plenty of time to consider their own decision.

Despite the simple nature of the question, over 35 percent of the people tested provided an answer they felt to be incorrect. This has nothing to do with visual impairment; in control experiments without plants, people chose correctly almost 100 percent of the time, and during the actual experimental sessions, test subjects would remark on how clearly wrong the other people in the room were.

Asch concluded that either the subjects didn't trust their own judgment when confronted with a number of opposing opinions, or they were uncomfortable voicing a conflicting opinion against a majority decision. He concluded that, for them, being accepted was more important than being correct. Crucially, if even one other person agreed with the subject, they were much more likely to make the right decision.

The experiment has been repeated since with similar results. In one version, 58 percent of students in a study agreed with the statement "the right of freedom of speech should be suspended when the Government feels threatened." When questioned individually, all of them disagreed.

One criticism often leveled against parapsychology as a science is the inconsistency with which its experiments can, or can't, be repeated. Parapsychologists sometimes attribute this to their test subjects' variable psi abilities—or lack of them—while skeptics and disbelievers argue that there must be something wrong with the experimental process.

To protect themselves against accusations of fraud, corruption and incompetence, serious parapsychologists employ extremely rigorous experimental protocols in their laboratory work. Knowing this, they've looked elsewhere for explanations to their often-contradictory results.

One possible answer to their problem lies in what's known either as the Experimenter, or Rosenthal Effect, after Robert Rosenthal, the Harvard psychologist who first studied it. In 1961, Rosenthal asked subjects to rate photographs of people. Half the experimenters were primed to expect high ratings from their subjects, the other half to expect lower ratings. Their findings matched their initial expectations. Next, Rosenthal's team applied their ideas to experimenters working with rats. Half were told that their rodents had been specially bred to speed through mazes, the other half that their rats were specifically bred to fail. In two separate experiments, Rosenthal found that those with higher expectations of their rats garnered better results.

In 1968 the experiment was taken into the classroom. Teachers were led to believe that their pupils—who had undergone standard intelligence tests –were broken into three bands of intellectual ability. Examined eight months later, when compared to a control group, those children who were expected to do better, did so. By 1988 there had been 464 similar studies carried out, all reaching similar conclusions. More recently the experiment has been expanded into areas like courtrooms, nursing homes and offices. Rosenthal himself sees nonverbal behavior—body language and general attitude—as the key.

Parapsychologists Richard Wiseman and Marilyn Schlitz applied the experiment to their own field twice, in the mid 1990s, while testing whether subjects could sense they were being stared at from behind. In both instances Wiseman, a psychologist who tends towards skepticism, and conveyed this in his manner, achieved no notable results with his subjects, while the warmer, more psi-positive Schlitz scored above chance. Other parapsychologists have wondered whether the unconscious psi-abilities of certain experimenters might also nudge their experiments towards a positive, or negative, outcome.

The psi-challenged disbelievers are, of course having none of it, arguing that such claims make a mockery of scientific method. Perhaps they'd feel differently if their teachers had been nicer to them.

In 1961, Adolf Eichmann, one of the key architects of Nazi Germany's Final Solution, was sentenced to death in Jerusalem. Throughout his trial, Eichmann had insisted that he was "only following orders."

Eichmann's prosecutors portrayed him as a sadistic monster, a simplistic characterization that didn't satisfy American psychologist Stanley Milgram. Seeking to explore the relationship between authority, obedience and personal morality, Milgram set up "a simple experiment" at Yale University.

Summoned to the laboratory and greeted by a scientist in a white lab coat, the subject—usually replying to an advertisement—was given the role of "teacher." Introduced to a "learner" in another room, the teacher watched them being strapped into a chair with an electrode attached to their wrist. Seated behind a screen in front of a large electroshock machine, the teacher then read out a list of words to which the learner responded with corresponding words that they were already supposed to have memorized.

If the learner's response was wrong, the teacher was to apply an electric shock to them by pressing one of thirty switches, labeled from "Slight Shock," through "Intense Shock" and ultimately, "Danger: Severe Shock." For each incorrect response the teacher was told to increase the voltage, resulting in grunts, screams and, finally, silence from the learner.

In fact the learner was an actor; the real subject of this experiment was the teacher. When the learner's screams and pleadings reached a certain intensity, the teacher often asked the scientist whether they should continue; some teachers would simply refuse to carry on. This was the crucial moment: the lab-coated scientist—the epitome of scientific authority—would now firmly insist that the session continued.

In a poll before the experiment began, psychologists predicted that only one in a thousand subjects would administer the strongest shocks. In fact, during the first round of experiments, conducted using Yale undergraduates as "teachers," 60 percent were fully obedient and took the sessions to their ultimate "lethal" conclusion. Subsequent experiments, carried out all over the world until 1985, achieved similar results. There were no significant differences between male and female subjects.

"For many people," wrote Milgram, "obedience is a deeply ingrained behavior... a potent impulse overriding training in ethics, sympathy, and moral conduct."

Perhaps because of the unflattering picture of humanity they present, the Milgram experiments have been heavily criticized by ethicists. Nonetheless, we cannot afford to ignore their message.

"**G**reat changes must be made in the American way of life... either we do nothing and allow a miserable and probably catastrophic future to overtake us, or we use our knowledge about human behavior to create a social environment in which we shall live productive and creative lives..."

So wrote the psychologist Burrhus Frederic Skinner in his 1948 utopian novel *Walden Two*, which describes a community formed around his own ideas about human behavior and social conditioning.

Skinner's theory of "radical behaviorism" considers all behavior in animals, including humans, to be determined by their environment and, complimenting this, their environment to be determined by their behavior. To explore these ideas, Skinner developed the Skinner Box, a cage containing a lever with which food can be released by the test subject, usually after performing a particular learned action.

In his most intriguing experiments, food was automatically delivered to pigeons in a cage at regular intervals. The birds began to associate the deliveries with whatever they were doing beforehand, subsequently continuing to perform these actions between each food drop, as if to summon more goodies. Skinner interpreted their behavior as superstitious and it's reminiscent of the ritual practices of the South Pacific "cargo cults."

While his promotion of social conditioning, as in his book *Beyond Freedom and Dignity* (1971), and his denial of the individual mind, (because it cannot be objectively measured through scientific observation), have seen him labeled a totalitarian, Skinner's intentions do seem to have been benign—for example he advocated only positive behavioral reinforcement. But his theories remain highly controversial.

When photographs of his infant daughter Deborah inside what looked like a human Skinner Box appeared in a 1945 issue of *Ladies' Home Journal,* horrified readers presumed he was experimenting on her as he had done his pigeons. In fact this was a temperature-controlled

playpen he had designed and hoped to market as the "Heir Conditioner" or "Aircrib." But in later years rumors spread that Deborah had become suicidal as a result of her mistreatment—rumors she was eventually forced to dispel herself in public.

A major figure in the understanding of human and animal behavior, Skinner remains a hero to the approximately 150 people who, over 30 years later, still live in communes—one in Mexico, the other in America—directly inspired by *Walden Two.*

aris, May 1736. In front of a startled audience that included several doctors and priests, Marie "The Salamander" Sonet laid herself out on metal stools over a large fire, expressing no pain for a full nine minutes.

Sonet was one of several hundred *convulsionaires*, for whom violent, ecstatic spasms of the body, coupled with feats of extreme endurance, became acts of religious devotion. The Jansenists, as they were also known, would be stabbed repeatedly, pounded and skewered by professional assistants before astonished audiences, some of whom were then moved to join the squirming throng themselves.

Some people can withstand what ought to be insufferable pain. During the Napoleonic Wars, the French General Jean Moreau calmly smoked a cigar as a surgeon amputated both his shattered legs; in May 2004, the 27-year-old climber Aron Ralston used a penknife to amputate his own arm after it was trapped by a boulder in Utah. Others undergo major surgery using hypnosis instead of anesthesia.

"There is little doubt that in the lower

classes the sensation of pain is felt in a much less degree than in those of a highly intellectual and nervous temperament," wrote one doctor in 1896. Our understanding of pain has advanced considerably since then, but despite his Victorian snobbery, a recent experiment reveals his observation to be surprisingly astute.

In 2003 a team, lead by Bob Coghill at the Wake Forest University in North Carolina, asked 17 volunteers to measure various levels of heat on a scale of one to ten. Initial results showed that one person's "one" was another's "ten." Next Coghill repeated the process while the subjects underwent a brain scan. Those most sensitive to pain registered activity in the brain's cerebral cortex, specifically the prefrontal cortex, which is associated with memory, emotion and attention, and the anterior cingulate cortex, known to be linked with pain. The brains of the hard nuts registered nothing in these areas, although the thalamus—which receives pain messages from the spinal cord and peripheral nerves—was active in all 17 volunteers.

Coghill's experiment shows that we all receive the same basic pain signal—it's the cerebral cortex that creates our individual responses to it. In theory, as the *convulsionaires* and others demonstrate, switching off pain should be no trickier than controlling our emotions.

Offers of volunteers for a special (and excruciatingly painful) *Far Out* experiment should be directed to this book's publishers.

I n January 1803, the body of murderer George Forster was pulled from the gallows of Newgate Prison and taken to the Royal College of Surgeons. Here, before a transfixed audience of doctors and curiosity-seekers, Giovanni Aldini, nephew of the famed electrical pioneer Luigi Galvani, prepared to return the corpse to life.

At least, this is what some of the spectators thought they were witnessing. When Aldini applied conducting rods, connected to a large battery, to Forster's face, "the jaw began to quiver, the adjoining muscles were horribly contorted, and the left eye actually opened." The climax of the performance came as Aldini probed the murderer's rectum, causing his clenched fist to punch the air, as if in fury, his legs to kick and his back to arch violently.

Aldini's was just one of many such experiments using human and animal corpses. He and the other "galvanists" were continuing the research of the late Galvani who, a decade previously, had demonstrated the effects of electrical current on frogs' nervous systems. In line with late eighteenth century "vitalist" ideas about a mysterious "life force," Galvani explained his findings by proposing the existence of "animal electricity." This electrical juice, he suggested, was generated in the brain, flowing through the nerves and supplying muscles with their power.

Although a great proponent of electricity's medical potential—it was used at the time to treat paralysis, rheumatism, as a purgative and to resuscitate drowned people, amongst other things—Aldini admitted that he was unable to restart a lifeless heart. Others were less modest, including Carl August Weinhold, a German scientist who claimed to have brought animals back from the dead. In a series of experiments, Weinhold extracted the spinal cords of decapitated kittens, replacing them with zinc and sliver pile batteries, which generated an electrical charge. Not only did their little hearts start beating again but, according to Weinhold, the kittens bounded around the room for several minutes. Weinhold would later propose enforced genital infibulation for all young men, an idea received with less enthusiasm than his prancing zombie kittens.

The electrifying demonstrations of Aldini, Weinhold and others contributed much to our understanding of physiology and electricity. But perhaps their greatest claim to fame is inspiring Mary Shelley's *Frankenstein* (1818), the book that forever shaped the image of the mad scientist in the popular imagination.

Y ou're walking down a lit street at night, deep in thought or otherwise distracted. Up ahead, a sodium street lamp casts an orange glow. As you approach, the light flickers for an instant, then switches off, only to come back on again once you've passed by.

Many of us will have experienced this at some time, and it's reported often enough for parapsychologists to have given it a name, Street Light Interference (SLI). Those who feel that they are repeatedly affected refer to themselves as SLIders.

The majority of SLIding incidents are easily explained. When common sodium and mercury vapor streetlights wear out they tend to "cycle"—flicker on and off—due either to changes in the consistency of the vapor inside them, or to aging electrodes. At other times the light-sensitive cells that operate them can develop glitches, leading to whole banks of lights behaving peculiarly. On witnessing this, our pattern-detecting minds make the connection and assume that we interfered somehow with the light's mechanism.

That, at least, is the rational answer, but naturally there are complications. Some SLIders claim to affect other electrical appliances, merely through proximity to them. Light bulbs, televisions, computers, watches, fridges, even the starters in their cars may malfunction, often permanently.

Sometimes referred to as High Voltage Syndrome (HVS), such cases have been recorded as far back as 1837, when a young American woman found herself dramatically charged with electricity for five months. Anyone she passed her hands over would be painfully zapped with static electricity. More recently in 1976, after breaking his arm, 12-year-old Vyvyan Jones of Bristol, England, found that he could illuminate light bulbs simply by touching them. TVs and lights would also flicker in his presence. In 1983, Jacqueline Priestman of Manchester, England, claimed to have blown up thirty vacuum cleaners, five irons and two washing machines. She also caused TVs to spontaneously change channel. Another Mancunian, Mandy Boardman, got through six vacuum cleaners and six TVs in three years. Her friends noted that she always gave them a shock when touching them. In September 2005 Frank Clewer of Warrnambool, South West Australia, generated 40,000 volts of static electricity simply by walking around the town in a nylon jacket and wool shirt. As he entered an office for a job interview there were loud crackling sounds before the carpet he was walking on burst into flame. Afterwards he also burnt some of the plastic in his car.

What causes people to retain so much static electricity remains unknown, but it's not necessarily a modern condition. Despite the massive increase in electricity use, recorded instances don't seem to have risen all that dramatically in the past 150 years. ⬚

081 MIND THE ZAP

Applying a tiny electric current to the front of the head can boost your verbal agility, according to scientists from the National Institute of Neurological Disorders and Stroke (NINDS) in Maryland.

Electrodes delivered two thousandth of an ampere—less current than is used by a digital watch—to the foreheads of 103 volunteers, who were then asked to name as many words as they could beginning with a specific letter. After 90 seconds, most people achieved 20 or so words but, after being zapped for 20 minutes, this rose by 20 percent. Control subjects who were wired up but not electrified scored only average results, as did those who received half the current.

Researchers suspect that the current affects cells in the brain's pre-frontal cortex, thought to be involved with aspects of the self as diverse as working memory, responsive movement and turning thoughts into words. Once the current has passed through them, the cells are able to activate more easily and so work more quickly. It's hoped that applying tiny currents in this way might help people with speech problems caused by damage to this part of the brain, including aphasias, in which people have trouble using or understanding words. In these disorders, the parts of the frontal cortices that are damaged can determine which types of words, for example verbs or joining words, are affected.

Electricity is now being used and investigated in pain control, treatment for depression, muscle control, even weight loss. The brain is essentially an electrical organ, its billions of neurons, or brain cells, are extremely sensitive to electrical signals fired off by their neighbors. Hence our sensitivity to even tiny electromagnetic fields. The abnormal firing of brain cells can cause migraines, facial tics, epilepsy and, on a larger scale, full-blown seizures.

Neurologists have been applying electromagnetic fields to human and animal brains, through the internal use of electrodes and, less controversially, external stimulation, since the 1950s. But electrotherapy goes back much farther; the ancient Greeks are known to have applied the torpedo fish (a form of ray) to the head and other parts of the body, to ease headaches and local pain, including hemorrhoids.

It won't be happening any time soon, but the NINDS team imagines electrical "thinking caps" assisting the brains of the future. And perhaps electric underwear will soothe their bottoms. ⌨

082 EROTOTOXIN

I t is all around us, seeping into our brains, and those of our children, via magazines, newspapers and television. Once there it gets to work, "reflexively and mechanically restructuring the brain." Even more terrifyingly, "involuntary cellular change takes place *even during sleep*, resisting *informed consent...*" by the subject.

According to Dr. Judith Reisman, pornography doesn't just affect your mind, but also the physical structure of your brain—turning you into a veritable porno-zombie. Porn, she says, is an "erototoxin," producing an addictive "drug cocktail" of testosterone, oxytocin, dopamine and serotonin with measurable organic effects on the brain.

Some of us might consider this a good thing. Not Reisman: erototoxins aren't about pleasure; they're a "fear-sex-shame-and-anger stimulant." Reisman's ability to produce copious amounts of foaming psychobabble—witness her paper "The Psychopharmacology of Pictorial Pornography Restructuring Brain, Mind & Memory & Subverting Freedom of Speech"—have made her the darling of the anti-pornography crusade, and in November 2004 she presented her erototoxin theory to a US Senate committee on porn's harmful effects.

Under the auspices of Utah's Lighted Candle Society (LCS), Reisman and Victor Cline, a clinical psychologist at the University of Utah, began raising money from American conservative and religious organizations. They hope to raise enough, at least $3 million, to conduct MRI scans on victims under the influence of porn and so prove their theories correct. They foresee two possible outcomes. If they can demonstrate that porn physically "damages" the brain, it might open the floodgates for "big tobacco"-style lawsuits against porn publishers and distributors. Second, and more insidiously, if porn can be shown to "subvert cognition" and affect the parts of the brain involved in reasoning and speech, then "these toxic media should be legally outlawed, as is all other toxic waste, and eliminated from our societal structure." What's more, people whose brains have been rotted by pornography are no longer expressing "free speech" and, for their own good, shouldn't be protected under the First Amendment of the American Constitution.

But there's a catch. Tragically, much of Reisman's research in developing her theory has necessitated examining hundreds, perhaps thousands, of pornographic magazines and films. By her own reasoning, her own brain ought, by now, to be a seething mass of toxic smut-mulch, horribly damaged beyond repair… Nurse, the restraints!

The Egyptians prescribed opium as a cure for insomnia over 3000 years ago, and Hippocrates recognized sleep's importance to a healthy mind and body in about 400 B.C.E., but the scientific quest for the active mechanism of sleep—what causes us to slip from wakefulness into slumber—really began in the nineteenth century. Two popular ideas were that sleep arose either from a lack of blood—and therefore oxygen—in the brain, or through a build up of chemical toxins such as lactic acid, carbon dioxide and cholesterol, that needed to be filtered and expelled from the body during sleep.

In 1907, French scientists Rene Legendre and Henri Pieron announced the results of a series of experiments on dogs. The hapless animals were kept awake for up to 10 days, tied to the wall by a collar so that they were unable to lie down and sleep. They were then killed and their cerebrospinal fluid extracted and injected into the nervous systems of healthy, active dogs. When these were seen to fall into a deep sleep after about an hour, the scientists regarded this as evidence of an active sleep-inducing molecule, which they called "hypnotoxin." Simultaneously in Japan, Kuniomi Ishimori was carrying out similar experiments and drawing the same conclusions, calling his sleep juice a "hypnogenic substance." While Legendre and Pieron's work is now famous, Ishimori's remained largely unknown in the West until recently.

Pieron's 1913 book *Le Probleme Physiologique du Sommeil* was the first to consider sleep from a physiological perspective, but the scientific establishment rejected his theories, regarding the hypnotoxin-induced sleep as closer to narcosis than natural slumber. His ideas were soon overtaken by those of Ivan Pavlov, who proposed that sleep resulted from the brain ceasing activity—an idea that had already gained favor in the nineteenth century. But hypnotoxin would return.

In 1967 an American scientist, John Pappenheimer, repeated Pieron and Legendre's work using goats, with the same results. This time they were able to isolate what they called "Factor S," and in 1982 this was identified as muramyl peptide, a molecule produced by the immune system. Research into its role in the sleep process continues to this day.

As the quest for both the perfect sleep inducer—and sleep inhibitor—continues, hypnotoxins may one day be big business. ◌

Of the billions of miles of DNA inside each one of us, about 95 percent is currently unaccounted for. This non-coding material, the dark matter of genetics, was prematurely labeled "junk DNA," with the implication that, because we didn't know what it did, it was of no use to us. This may have been one of the costliest examples of scientific arrogance in recent history.

For some time junk DNA, which exists in differing proportions in all living species, was effectively ignored, but when, in the early 1990s, it was examined using methods of linguistic analysis, it revealed patterns similar to those found in ordinary language. Other tests identified areas of symmetry, likened to linguistic palindromes, which further suggested an underlying structure.

A number of ideas have been mooted for the function of this unidentified DNA. Much of it is thought to consist of pseudogenes, "molecular fossils" no longer required for an organism's evolutionary development. Some have suggested that these pseudogenes remain in "standby" mode until alterations in the organism's environment once again necessitate their use or, perhaps, further evolutionary changes.

It's now believed that junk DNA could yield vital clues to the genetic mapping, and even prediction, of many diseases, including several cancers. But to the dismay of genetic ethicists, these clues will probably remain the property of an Australian Biotech firm, Genetic Technologies Ltd. (GTG).

In the late 1980s, immunologist Dr. Malcolm Simons began to wonder why, if junk DNA is useless, all living beings contain so much of it. Suspecting that the patterns within it pointed to some as-yet-unknown function, Simons and businessman Mervyn Jacobson filed two patents for mapping and analyzing non-coding DNA across all living species. To the surprise of many, the patents were approved in over 20 countries.

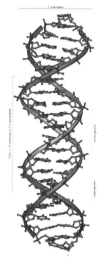

As our understanding of junk DNA grows, so the decision to grant Simons these patents—and GTG's decision to charge academic institutions for the use of their patented technology—have come increasingly under fire. The Australian geneticist John Mattick appears to speak for many when he says "the failure to recognize the implications of the non-coding DNA will go down as the biggest mistake in the history of molecular biology."

I n 1985, Swiss-Canadian anthropology student Jeremy Narby spent a year at Quirishari in the Peruvian Amazon, studying how the Ashaninca tribe made use of indigenous resources.

Wondering where the Ashaninca's extensive knowledge of the jungle's plant and animal life originated, Narby was always pointed towards the *ayahuasceros*—shamans whose work is carried out under the influence of the hallucinogenic plant brew ayahuasca. The shamans told Narby that their knowledge came to them during ayahuasca sessions, and that they were taught by nature itself.

This mystical response initially puzzled Narby. It seemed to conflict with the Ashaninca's necessarily pragmatic relationship to

their challenging jungle habitat. But this is just one of the many puzzles surrounding these and other Amazon-dwelling peoples, and their innate understanding of nature's pharmacopeia.

Out of more than 80,000 plant species found in Amazonia, how did early shamans stumble upon this incredibly potent brew? Ayahuasca is composed of two key ingredients: the psychoactive *dimethyltryptamine* (DMT), found in several jungle plants and trees, and the

Banisteriopsis caapi vine—considered the "soul" of the brew by aya-huasceros. Chemical inhibitors in the vine prevent the DMT from being rendered inactive by enzymes in the stomach.

It seemed remarkable to Narby that the shamans had discovered such a chemically complex mixture purely through trial and error; and yet it's just one of hundreds of plant and animal-based preparations used by the tribes, many of which are only now being investigated by pharmaceutical science and industry.

The Ashaninca ayahuasceros told Narby that they spoke to serpents after drinking their magical brew, while his own ayahuasca experience saw him having a perfectly sensible conversation with a huge snake. Sifting through records of indigenous shamanic experiences and cre-ation myths from Australia to Scandinavia, Narby was amazed by how many of them featured imagery of twisted vines, rope ladders, cre-ator serpents and twins—forms he found deeply suggestive of the twin double helix of DNA.

Finding that DNA emits extremely weak, but brightly-colored bio-photons, Narby then suggested that these might form the basis for the luminous patterns in the ayahuasceros' visions. So are the shamans gaining their information from DNA itself?

Narby thinks it's possible and outlined his theory in the book *The Cosmic Serpent: DNA and the Origins of Knowledge* (1998). Although such grand unified theorizing is considered anachronistic in contemporary anthropology, Ayahuasca use has now spread far from its Amazon origins and Narby's ideas have gained an enthusiastic following amongst anthropologists, ecologists and psychonauts the world over. ◻

086 TUSKO'S LAST TRIP

The annals of science describe thousands of noble experiments monitoring the effects of recreational drugs, from alcohol to tetrahydrocannabinol (cannabis extract), on animals including monkeys, dolphins, pigeons and spiders. But the biggest and most controversial animal drug experiment involved a three-ton bull Asian elephant named Tusko.

Conducted by Dr. Louis Jolyon "Jolly" West and two colleagues, the experiment took place in 1962 at the University of Oklahoma. West's stated intention was to see whether LSD—yet to hit the streets as a recreational drug—would induce a condition called *musth* in Tusko. *Musth*, which occurs naturally in all bull elephants, is a period of heightened testosterone production and high aggression. Why West should be interested in this is unclear, though he has been repeatedly linked to the CIA's MK ULTRA program, which had been experimenting with LSD on unwitting subjects like Tusko since 1953. One likely possibility is that West was interested to see if LSD could heighten aggression in human soldiers on the battlefield.

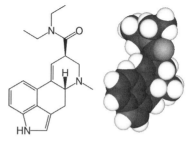

Tusko, "the prize of Oklahoma City Zoo," was injected with 297 mg of LSD—an enormous dose, even for an elephant, and over thirty times what a three-ton human should receive. After five minutes, Tusko trumpeted, fell over, defecated and began shuddering violently; his pupils dilated, his legs became stiff, he bit his tongue and his breathing became extremely labored.

Twenty minutes later, in an attempt to calm him down, a large (again, almost certainly too much) amount of the anti-psychotic Thorazine was injected into the elephant, likely inducing a massive drop in blood pressure and heart palpitations. But it didn't help; after another hour West pumped Tusko with a tranquilizer, and a few minutes later he was dead. The whole process took one hour and 40 minutes.

A great deal of controversy surrounds the Tusko experiment. Rumors persist that West was on LSD himself during both the experiment and the autopsy afterwards, and that he also shot Tusko up with amphetamines. While the experiment is quoted as evidence of LSD's toxic nature, it seems most likely that the Thorazine, or the combination of drugs, killed Tusko, rather than the acid. Lending credence to this, in 1984 the psychologist Ronald K. Siegel repeated the experiment with two elephants, using LSD only—in appropriate, though elephantine, doses—and both survived.

Tusko's is one great leap for elephant-kind that need never be made again. 🖵

087 KINDS OF BLUE

The early nineteenth century saw the first appearance of hashish and cannabis in Western Europe, largely as a side effect of Napoleon's failed invasion of Egypt. Curious about its potential for easing mental illness, the psychiatric pioneer Dr. Jacques-Joseph Moreau De Tour began to experiment with the drug, both on himself and on his patients at the Bicêtre mental hospital outside Paris.

Moreau's own experiences with hashish, consumed Arabic-style in a paste called *dawamesc*, were astoundingly intense and resoundingly hilarious, convincing him that he was on the right track in identifying either the roots of madness, or its cure. But he needed more test subjects. So, like Humphrey Davy testing nitrous oxide on the Romantic poets before him, and Timothy Leary with LSD a century later, Moreau turned to the *litterateurs* for inspiration.

The Club des Hashishins, founded by Moreau and the romantic poet and writer Théophile Gautier in 1844, assembled monthly under the baroque finery of Paris' Hotel Lauzun. There its members, including Alexandre Dumas, Gerard de Nerval, Honore de Balzac and Charles Baudelaire, consumed *dawamesc* in varying dosages, recording their experiences for Moreau and for posterity.

Moreau was particularly interested in how dosage altered the users' experiences. Small amounts of hashish often provoked "paroxysms of mirth"; increased doses led to time loss and dissociation, while high doses caused the psychonauts to be engulfed by great blue clouds of transcendent awareness, feelings memorably described in Baudelaire's *Poem of Hashish*.

Animal experiments echo Moreau's findings. Psychopharmacologist Dr. Ronald Siegel gave cannabis extracts to Rhesus monkeys trained to recognize the difference between a real chocolate sweet and a holographic projection. Low doses produced confusion and frustration in the choc-hungry monkeys. At increased doses they lost interest in the sweet itself and began to play with the chocolate mirage instead. Even higher doses resulted in "girning"—monkey laughter—while extreme doses induced a state of languid, apparently contented, simian contemplation.

Siegel then trained the same monkeys, as well as pigeons and human subjects, to identify colors, shapes and complex patterns projected onto a screen, pressing buttons to identify what they were seeing. Given cannabis resin and a blank screen, all three species—who share similar perceptual systems—described the same thing: shifting, blue geometric patterns, like the clouds described by Moreau's *hashishin*. 〇

088 RAT HEAVEN

D o depressed lab rats dictate international drug policy?

The predominant model of drug addiction views it as a disease: humans and animals will use drugs like heroin or cocaine for as long as they are available. When they run out they will seek a fresh supply and continue: the drugs, not the users, are in control.

These conclusions, repeated regularly by politicians and the media, are based on experiments carried out almost exclusively on animals, usually rats and monkeys, housed in stark metal cages and experiencing a particularly poor quality of life. What would happen, wondered psychologist Dr. Bruce Alexander, then of British Columbia's Simon Fraser University, if these animals were instead provided with a comfortable, stimulating environment?

In 1981 Alexander built a 656 square ft (61 m^2) home for lab rats. Dubbed Rat Park, it was kept clean and temperate, while the rats were supplied with plenty of food and toys, along with places to dig, rest and mate. Alexander even painted the walls with a soothing natural backdrop of lakes and trees. He then installed two drips, one containing a morphine solution, the other plain water. This was rat heaven: but would happy rats develop morphine habits?

Try as he might, Alexander could not make junkies out of his rats. Even after being force-fed morphine for two months, when given the option, they chose plain water, despite experiencing mild withdrawal symptoms. He even laced the morphine with sugar, but still they ignored it. Only when he added Naloxone, an opiate inhibitor, to the sugared morphine water, did the Rat Parkers drink it. Alexander simultaneously monitored rats kept in "normal" laboratory conditions (i.e. small cages): they consistently chose the morphine drip over plain water, sometimes consuming 16–20 times more than their luckier colleagues.

Alexander's findings—that deprived rats seek solace in opiates, while contented rats avoid them—dramatically contradict our currently held beliefs about addiction. So, how might society benefit if his results were applied to human drug addicts? Nobody seemed to care. Rejected by *Science* and *Nature*, Alexander's paper was published in the obscure *Pharmacology, Biochemistry and Behavior*, where it was summarily ignored.

Two decades later Rat Park sits empty; addiction remains a "disease" and the ludicrous War on Drugs continues, with no end in sight. ⌑

PART FIVE

FAR OUT!

B ig Paddy and Harry were aboriginal trackers working closely with the Mounted Police in Derby, West Australia. In February 1954, the pair stumbled upon three aborigines wanted for murder. Two fled, but the other, a sorcerer, pointed a bone—an Ungulla, or poison stick—at them and began screaming death curses. Within days both trackers were dead.

While working as a teacher in Hawaii in the early twentieth century, the psychologist Max Freedom Long became fascinated with Polynesian witchcraft. Islanders particularly feared Ana Ana, the death prayer, by which a sorcerer could induce another person's death, usually in the name of justice. It's said that its victims all died in the same way, of a slow, creeping paralysis that first affected the lower extremities, gradually rising through the body until it reached the lungs, at which point the victim died of respiratory failure. Today, doctors would probably identify this as Guillain-Barre syndrome, an illness clinically identical to the symptoms of the death prayer. For those infected in the past, the chances of survival were slim, though these days patients are kept alive on ventilators until the disease subsides, which can take weeks.

So can curses really kill? A victim could theoretically be scared to death: fear stimulates adrenaline flow, making the heart rate increase and breathing become shallow. If prolonged, this could lead to a fatal heart attack. Reports throughout history have also described people willing themselves to death; becoming convinced that they will die on a certain day, even if they are otherwise healthy.

This might explain cases in which people knew that they had been cursed, but what about those in which they didn't? Psychics involved with the American military have claimed that they were ordered to use their powers to kill, but for obvious reasons no scientists have openly studied the efficacy of death curses. Several experiments, however, have been carried out into the healing power of prayer, with mixed results.

In 2000, Professor Leslie Francis of the University of Bangor surveyed several prayer experiments, reaching the conclusion that prayer worked both for those praying for themselves, and for those being prayed for by others. But another experiment, headed by Dr. Mitch Krucoff at Duke University Medical Center in 2003, found otherwise. Names from a sample of 750 heart patients in nine hospitals were randomly sent to 12 prayer groups worldwide, incorporating Christians, Buddhists and Muslims. Krucoff found that the patients who had been prayed for healed no more quickly that those who hadn't.

Perhaps sealing the issue, in April 2006, Harvard University Medical School cardiologist Herbert Benson, conducted the largest investigation into the power of prayer yet, with a total of 1802 patients from six US hospitals. Those doing the praying began the night before the patient's surgery, and continued for two weeks afterwards. This time, there were no significant differences in recovery rates between those who knew they were being prayed for, those who didn't or, indeed, those who weren't being helped at all. While it may have produced some ease of mind for the patients who did know they were being prayed for, it didn't actually aid their condition.

Prayer, it seems, just doesn't work and, if we can assume that prayers for death are no more potent than prayers for health, Polynesia's sorcerers may need to take up more conventional weapons if they want to stay in business. ⌐

090 WATER WITCHING

"Call me old-fashioned, but I rely on my rods," engineer Dougie Scriven told journalists as he retired from the UK's Yorkshire Water in 2001. "I've used them for 24 years now and they have come up trumps when everything else has failed." Before leaving, Scriven trained several new employees how to use dowsing rods, just as he was taught to use them on joining the company in the 1970s.

Divination for water, oil, minerals and other objects—traditionally using a forked twig of hazel or yew, copper wire, a pendulum or even a bent coat hanger—is still practiced all over the world. Known as water witching, radiesthesia, rhabdomancy, divining or dowsing, it's a skill that has been practiced for thousands of years across every continent.

EPREUVE par la BAGUETTE.

There are innumerable historical accounts of dowsing successes, from the 1692 case of Jacques Aymar in Lyon, France who used divining rods to track down three murderers, to its use in seeking tunnels and traps by US engineers in the Vietnam War. Today dowsers are still as much a part of our rural existence as crows and tractors.

The forces behind the ability have been ascribed to everything from electromagnetism and radiation to subconscious readings of the landscape or human body language, but ultimately they remain unidentified. They are also temperamental. Typically, when studied under controlled conditions, the dowsers seem unable to perform as well as they had expected, often scoring results close to chance. As yet, no dowser has claimed the million dollars proffered by arch-debunker James Randi to anyone satisfactorily demonstrating paranormal abilities.

One famous study, carried out near Munich, Germany in the late '80s, saw 500 dowsers perform almost 10,000 double blind trials detecting pipes buried underground. The physics professor behind the experiment, Hans-Dieter Betz of Munich University, declared that he had incontrovertible evidence of the dowsers' abilities, though hard line statisticians have since called the findings into serious question.

But for every doubting scientist there's a success story. In 2003, a dowser hired by the Ysgol Gyfun Preseli school in Pembrokeshire, Wales located an underground water supply which, it is hoped, could save them £10,000 ($19,600) a year in water bills.

091 FIRESTARTERS

Sydney, Australia, August 1998; 60 °F (15.5 °C). Jackie Park picked up her mother Agnes Philips, an Alzheimer's sufferer, from her nursing home and dropped into a shop, leaving Agnes asleep in the car. Minutes later Jackie saw smoke coming from the vehicle, then flames. Rushing out, she saw her mother being dragged out by a passerby, repeating calmly "It's too hot" as her burning body was extinguished.

Agnes died a week later. At her inquest, the fire inspector was unable to determine what had started the blaze; the car's engine was not running and no fuel or exposed wiring was found anywhere. An open verdict was recorded.

As a possible instance of Spontaneous Human Combustion (SHC), Agnes' mysterious demise is unusual, though not unique, for having been witnessed by others, and her fires being put out before she was entirely consumed by flames. More usual is the 1999 case of an octogenarian woman whose charred torso—and nothing else—was found in her otherwise undamaged Paris flat.

Documented for centuries, SHC was once considered an occupational hazard for heavy drinkers, a fate employed by Dickens for Krook's death in *Bleak House*; but we may now be closer to understanding the phenomenon. A 1998 investigation by the BBC's *QED* TV program used a dead pig to demonstrate something known as the "wick effect." Wrapped in clothes that act as a wick, the fat surrounding a person burns slowly but fiercely, like a human candle. Given long enough—at least five hours—even bones will be reduced to ash in this way, something which crematoria, designed to work fast at temperatures of around 1400 °F (760 °C), often cannot manage. The candle effect also demonstrates why it is usually only the extremities, often the least fatty parts of the body, that remain.

This theory can account for several SHC cases, but there are problems. The wick effect requires both an initial naked spark or flame (which cannot be identified in many alleged SHC incidents) and a subject who, for some reason, does not try to escape. Some victims have remained seated throughout their ordeals; they were presumably quickly asphyxiated, or incapacitated by sleeping pills or alcohol—suggesting that perhaps drunkards are more susceptible to SHC after all. ▢

092 STRANGE ATTRACTORS

Since 1987, one year after the Chernobyl disaster, 76-year-old Russian factory worker Leonid Tenkaev, his wife Galina, their daughter Tanya and grandson Kolya have all been able to make metal objects stick to their bodies. Leonid himself can hold individual objects weighing up to 50 lb (23 kg) on his chest.

Doctors in Russia and Japan appear convinced that the Tenkaev's abilities are genuine, describing, for example, how difficult it is to pull ferrous objects away from their bodies. "There is absolutely no doubt that the objects stick as if their bodies were magnetic," an impressed Dr. Atusi Kono told reporters in 1991.

Remarkably, the Tenkaevs are not alone. In 1990, the Superfields conference in Sofia, Bulgaria attracted 300 such "human magnets" after a young woman, Marinela Brankova, demonstrated her powers on television by supporting a 15 lb (7 kg) weight from her vertically held palms.

Within the past decade it has been shown that magnetic particles do exist in the human brain, but only in minute quantities. There seems to be no connection between these and the phenomenon at hand, which, if genuine, appears to be a form of telekinesis. Although people with the gift are usually referred to as magnets, many of them can also hold plastic, glass, wood and paper objects, with some stating preferences for specific materials. Nor does it seem to be an electrostatic phenomenon: subjects at the Superfields conference were able to demonstrate this by attracting items through thick rubber gloves.

The adhesive force seems predominantly to affect the upper body: the chest, arms and hands. Practitioners state that it can be fortified through practice and increased concentration—some people, while supporting several objects at once, can release specific items to order. One Bulgarian woman, Victoria Petrova, entertained delegates by making objects move about her body in time to music.

Some human magnets also claim other abilities, such as X ray vision similar to that displayed recently by the Russian, Natalia Demkina. Curiously there does seem to be a preponderance of (or perhaps just a preponderance of interest in) such powers in Russia, Bulgaria and other Eastern European countries, leading some researchers to connect it to radiation leakages. However, reports of human magnets from at least the mid-nineteenth century would suggest that its origins lie elsewhere.

At the 2003 British Association Festival of Science, the late Professor Robert Morris, of Edinburgh University's Koestler Parapsychology Unit, announced that his team's experiments continue to suggest the reality of telepathy.

While Morris avoided the T word, preferring the broader term "anomalous cognition," his team's research is merely the tip of a very ancient iceberg. Herodatus recorded the first known telepathy experiment when, in 550 B.C.E., King Croesus of Lydia challenged seven famed oracles to tell his messengers exactly what he was doing on a given day. Only the Pythia, the priestess of Apollo at Delphi, answered correctly—Croesus was making lamb and turtle stew in a bronze kettle. The tale may be apocryphal, and Croesus' misinterpretation of the Oracle's advice eventually led to his defeat by the Persians; but as an experiment, parapsychologists admit it wasn't bad.

Modern interest in thought transference arose in late eighteenth century France, when it was observed as a side effect of Franz Mesmer's proto-hypnotic practices. The term telepathy itself—meaning distant occurrence or feeling—was coined by Frederic Myers, a founder of the Society for Psychical Research, in 1882. With the respectability granted by SPR's scientific investigations, telepathy soon became a hot topic of *fin de siècle* salon culture, perplexing great minds from Oscar Wilde to Sigmund Freud.

In the 1920s and '30s, J. B. Rhine's experiments using Karl Zener's symbol cards captured the public imagination and popularized the notion of Extra Sensory Perception (ESP). Rhine's dry statistics—still a hallmark of the science of parapsychology—were supplemented by the conviction expressed in Pulitzer-prize winner Upton Sinclair's book *Mental Radio* (1930), in which he detailed several successful experiments he conducted at home with his wife.

Today's telepathy experiments, including those of Morris' team, tend to use the Ganzfeld method, in which the subject's ordinary audio and visual senses are suppressed, via white noise in the ears and half ping pong balls taped over the eyes. A "sender" then views images or films and attempts to transmit their impressions to the receptive subject.

Ongoing experimentation points to a number of factors that might increase a subject's telepathic hit rate, including a pre-existing belief in psi phenomena, a relaxed demeanor and, perhaps more worryingly, scoring highly on the Schizotypal Personality Disorder test—i.e. being a little odd.

Ironically, just as appearances like Morris' at science festivals would seem to represent a growing acceptance of psi within the mainstream, so a number of key parapsychology research units—including the Koestler unit that he headed—are closing down. With their emphasis on statistics, dramatic, media friendly results rarely emerge from the parapsychologists' laboratories, but this may be what's required to reverse this disappointing trend. ♇

094 THE REMOTE VIEWERS

n September 1995, the US military simultaneously publicly revealed and closed down Project Stargate, a 23-year, $20 million program of psychic spying, or Remote Viewing.

The project began in 1972 after the CIA became concerned by reports that the Soviet Union was dedicating substantial resources to what it called "psychotronics"—research into potential military applications of several psychical and fringe science phenomena.

In order to close this Cold War "psychic warfare gap," the CIA set out to assess how serious Soviet psychotronic threat really was. In 1972 they contacted parapsychologists Hal Puthoff and Russell Targ of the respected Stanford Research Institute, asking them to look for repeatable psychic phenomena that might be militarily useful. Working with talented psychic and artist Ingo Swann, Puthoff and Targ developed what they called "A Perceptual Channel Across

Kilometer Distances," i.e. the ability to witness objects, people and events at a distance: Remote Viewing.

The initial project—called Scanate, meaning "Scan by Co-ordinate"—required the viewer to describe what they could see at specific grid references provided by the CIA. It was deemed successful enough to convince the government to expand the project. Undergoing several name changes—including Sun Streak, Grill Flame, and, finally, Stargate—the RV program was involved in assisting hundreds of US military and intelligence-gathering operations over its 23 years. It would score some notable successes, and plenty of failures.

The team is said to have identified spies (in 1980, a KGB agent in South Africa using a pocket calculator to transmit information); located new Soviet weapons and technologies including a large nuclear submarine in 1979; helped to find lost SCUD missiles in the first Gulf War and sought plutonium in North Korea in 1994. Over the years more than twenty psychics were employed. It was grueling work—some of them ended up recovering in psychiatric hospitals, others lost the plot and became obsessed with crop circles and aliens.

Although $20 million is peanuts by US military standards, the project was closed down in 1995 largely because, it is suspected, the Defense Department was embarrassed by it. Some of its psychics continue to do government work; one assisted the FBI—apparently unsuccessfully—during the hunt for Osama bin Laden in late 2001.

It's not just a US military enthusiasm—in 2002 Britain's Ministry of Defence tried to "remotely view" bin Laden, also unsuccessfully. The budget was £18,000 ($35,300), perhaps reflecting the UK's much smaller military budgets. 🗨

095 THE VANISHING

A staple of '60s and '70s TV sci-fi and the dream of courier companies and military strategists everywhere, teleportation has its roots in the mystical flights of saints, documented since at least the sixteenth century as "transvection," and in nineteenth century séance rooms, where objects known as "apports" would mysteriously appear from the aether.

The word itself was coined by Charles Fort, the early twentieth century compiler of anomalies, in his 1931 book *Lo!*. Fort imagined teleportation as a force of nature, an "organic intelligence" that at one time helped to seed life across the planet. It could also be harnessed, both intentionally and unintentionally, by humans. But, writes Fort, "I'd ask whether something that mysteriously appears somewhere had not mysteriously disappeared from somewhere else."

Fort wasn't just referring to fish and frogs. There are numerous accounts of human teleportation, though few, if any, stand up to scrutiny. The most impressive is that of the seventeenth century nun Sister Mary of Agreda, "The Blue Nun" (so-called for her blue habit rather than any un-nunly behavior) who claimed to have been whisked from her native Spain to New Mexico over 500 times between 1620 and 1631. Legend has it that when Spanish explorers first reached the Indian tribes of this region in the early seventeenth century, they spoke of a pale "blue lady" who had taught them about Jesus and the sign of the cross.

Equally unlikely, though more entertaining, is the tale of Agnes Guppy. On the evening of June 3, 1871, Mrs. Guppy was at home with a friend in Highgate, London. Meanwhile, on Lambs Conduit St. in Holborn, about three miles away, a séance group was communicating with the famed spirit Katie King. A joker among the group asked Ms. King to bring them Mrs. Guppy, herself a medium and renowned as much for her ample girth as her ability to produce apports, to the sitting. Three minutes later there was a loud thump as something heavy dropped onto the séance room table. It was Mrs. Guppy, unconscious and dressed only in a nightgown, swiftly followed by some more respectable clothing for her return journey.

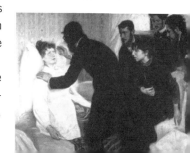

Mrs. Guppy's trip was the talk of the town, and so was she. Caught throwing apports around several times during séances, she was later accused of plotting to throw vitriol onto the face of a more popular rival medium, Florence Cook, Katie King's main channel, during another sitting.

More recently, teleportation was a key element of the late twentieth century alien abduction syndrome, during which people would be "beamed" in and out of their beds by extraterrestrials. But it seems that the aliens have yet to perfect their technology: some of their victims being deposited miles from home wearing only their underwear and, on at least one occasion, somebody else's entirely.

The US military has also taken a keen interest in teleportation.

"The psychic aspect of teleportation... has been scientifically investigated and separately documented by the Department of Defense." So begins an August 2004 paper, *Teleportation Physics Study* by Dr. Eric W. Davis, commissioned by America's Air Force Research Laboratory.

Part of the paper describes the theoretical manipulation of spacetime to create stargates or wormholes: "a traveler stepping through the [wormhole's] throat will simply be teleported into the other remote spacetime region or another universe," states Davis. Another proposed teleportation method employs quantum entanglement, what Einstein called "spooky action at a distance." Separated by any distance in space or time (forwards or backwards), particles that have been in contact with each other remain "entangled," so that whatever affects one particle also happens to the other. This effect has been replicated in laboratories, including the recent teleportation of a laser beam carrying a radio signal.

While these approaches work in theory, and in practice on a sub-atomic scale, they're still a few hundred years from becoming useful reality. But why wait centuries to harness unstable and potentially catastrophic cosmic forces when teleportation may already have been successfully demonstrated, as in the case of Mrs. Guppy? If we're honest, most nineteenth century spiritualist accounts have an anecdotal air about them. But according to Chinese lab reports analyzed by the US Defense Department in the 1980s, a number of carefully controlled experiments showed that psychics could teleport solid objects, including living insects, using only their minds.

A September 1981 paper in China's *Nature Journal* soberly described how children were able to teleport radio transmitters, horseflies, watches and other objects several meters, without touching them. The process lasted from under a second to a few minutes. Further research took place at Beijing's Aerospace Medicine Engineering Institute and these were filmed and photographed. The footage revealed that some teleported objects actually merged temporarily with the walls of their containers, while others would simply vanish, reappearing fully formed in a new destination. The signals from the teleported radio transmitters weakened or vanished entirely during transit, as if they were far away or, suggests Davis, in another dimension.

However, these reports may not be quite what they seem. During the Cold War, both sides generated reams of false information to confuse and frighten their enemies—tales such these may well prove too good to be true. □

096 TUNING IN TO THE GLOBAL MIND

A notion that has influenced esoteric thought for centuries now forms the basis of an ongoing, mind-boggling parapsychology experiment. Could our thoughts and intentions—before they become actions—affect the world around us?

During an EEG (electroencephalogram) test, electrodes detect the electrical signals transmitted between brain cells and record patterns of activity. This is not a measure of Mind itself, but of the complex electrical processes that somehow generate consciousness.

Now, imagine the Earth as a brain; us humans—perhaps all life—as brain cells; and a network of Random Event Generators (REGs—like high-speed, electronic coin flippers) as electrodes. This is the Global Consciousness Project (GCP), and it appears to be measuring, well... something. Begun in 1998, it now involves over 75 networked computers, known as EGGs ("electrogaiagrams"), in about 30 countries, including the USA, the UK, Russia, Fiji, Cuba and Romania.

The project grew from experiments carried out by Dr. Roger Nelson of Princeton Engineering Anomalies Research, one of the world's most prestigious parapsychology institutes. Since 1979, PEAR has studied the effects of human consciousness on REGs, demonstrating, to their satisfaction, that individual minds can subtly influence random mechanical processes, somehow creating deviations from the expected chance results.

Taking the idea further, Nelson examined what happened to a REG when several people focused on a single event, at a theatre or a sports stadium. The results were impressive but, perplexingly, he then found that the REG's physical location was irrelevant—the effects were present anywhere. REGs in the United States, for instance, were noticeably affected by Princess Diana's funeral in 1997.

Nelson's team claims that periods of widespread attention or concentration correspond to notable fluctuations in the EGG network's data. For example, significant results were recorded following the Turkish earthquakes of August 1999, on Millennium Eve, on the night of the 2000 US presidential elections and, of course, on September 11, 2001, when the GCP network responded in a "powerful and evocative way."

The GCP team is well aware that it is pursuing an extremely subtle beast, and remains cautious about interpreting its results. But the implications are startling and, as the EGG network continues to grow, so too does the enigma surrounding its data.

Sadly the GCP will now have to find a new home. In February 2007 PEAR was shut down by Princeton University. Disappointing as this may be, however, it seems unlikely that it will halt the GCP's network of enthusiastic researchers. ␡

097 THE WEIGHT OF THE SOUL

The question of Soul has entertained philosophers, thinkers and musicians since humans first became conscious of their selves.

For Aristotle and the ancient Egyptians the soul resided in the heart; for some Romans and Hebrews it existed in the blood; Plato and the ancient Greek physician Galen broke it into three parts, while some African tribes found it in the liver of animals, which they ate to enlarge their own souls. By the sixteenth century the idea that the mind, and therefore the soul, was housed in the brain, became increasingly standardized in medicine, though a hundred years later Descartes would propose his crucial split between mind and body, a split that many have been trying to heal ever since.

But can the soul's existence be proven scientifically? Dr. Duncan MacDougall of Haverhill, Massachusetts thought so, and set out to demonstrate as much to the medical community. He reasoned that if the soul was a real and measurable component of the human being, then at the moment of death, as the soul leaves the body, there ought to be a measurable change in the corpse's weight.

In his office, MacDougall built a bed, perfectly balanced on scales. He then called on six willing subjects in the late stages of terminal illnesses—patients "dying with a disease that produces great exhaustion, the death occurring with little or no muscular movement," including tuberculosis and diabetes, and, one by one, weighed them as they died. "The patient's comfort was looked after in every way," he insists in his paper "The Soul: Hypothesis Concerning Soul Substance Together with Experimental Evidence of The Existence of Such Substance," published in *American Medicine* in 1907.

MacDougall's first patient died after three hours, forty minutes. In this time he lost weight at the rate of one ounce per hour, shed through sweat and breath moisture. At the moment of death the bed suddenly dropped down, hitting a support bar with a clunk. The patient had lost three quarters of an ounce—21 grams. This, then, was the weight of the soul, an idea that was gladly picked up by the *New York Times* and became hard fact for Christians, Spiritualists and those who saw the 2003 film *21 Grams*.

But this doesn't take into account the results of MacDougall's other five subjects. Two lost over an ounce and a half between death and the few minutes immediately following it, two had to be rejected after problems with the apparatus and interruptions by "people opposed to our work," and one lost weight only to mysteriously gain it again. Weighing the soul was hardly an exact science.

Similar experiments with 15 dogs, proved, at least to MacDougall's satisfaction, that they didn't have souls, but he admitted that his human experiments would need to be repeated several times more before any definite results could be announced. MacDougall's work was roundly criticized at the time; both for its methodology, its dodgy ethics and his own general lack of understanding of what happens when we die.

Although he would experiment with the recently dead again, this time trying to catch the soul on camera, there would be no further research into this weighty problem. ▢

098 EXPERIMENTS WITH TIME

I n the years preceding World War I, Lieutenant John William Dunne was a famed aircraft designer, engineer and pilot. However, concurrent with his explorations of space, Dunne was making bold incursions into another dimension: time.

Beginning in 1899, Dunne noticed apparently meaningless coincidences spanning his dreaming and waking lives; for example, one morning he woke to find his pocket watch had stopped at exactly the same time as he had just dreamed. On studying his dreams

Biplan NIEUPORT DUNNE
14 m. d'envergure - Stabilisation automatique par Ailerons - Direction Ailerons
Moteur Gnome 80 H.P. - Piloté par le Capitaine FELIX

more closely, he came to believe that they somehow prefigured his knowledge of events in the waking world.

In May 1902, while in Africa, Dunne dreamed vividly of being on an island whose volcano threatened to erupt. He struggled to evacuate the island's 4000 inhabitants and narrowly escaped on a ship, from which he witnessed the catastrophic eruption. A few days later Dunne received *The Daily Telegraph*, which reported a volcanic eruption on the Caribbean island of Martinique with the loss of 40,000 lives—a figure he initially misread as 4000. Dunne wondered whether his dream had put him into telepathic contact with a survivor of the disaster or the article's author, or whether—with its simple numerical error—it was actually a premonition of his misreading the newspaper.

More such dreams followed. Some were mundane and personal, others were of air crashes (perhaps not uncommon for a pilot) and larger disasters. Noticing that details were often altered in the dreams—as they are in those based on real memories—Dunne came to believe that, rather than premonitions, these were ordinary dreams of events he would soon live or learn about, only somehow displaced in time.

Dunne's popular 1927 book, *An Experiment with Time*, inspired such influential writers and thinkers as J. R. R. Tolkien, J. B. Priestly and Buckminster Fuller. In it he portrayed time as a piano keyboard: in waking life we play one note at a time, from left (past) to right (future). Sleep frees us from this sequence and we can play the keyboard in a more abstract, non-linear fashion, memories from the future bubbling through in our dreams. He recommended that we keep a dream journal to identify such moments of temporal displacement.

And the experiment continues: in a 1997 paper, Mary S. Stowell examined 51 dreams from 5 subjects claiming precognitive abilities; of these, 37 proved accurate. Tales of everyday dream precognition crop up regularly in the news: in January 2006, English soccer fan Adrian Hayward dreamed that Liverpool player Xabi Alonso would score a goal from his team's own half. He bet £200 ($392), at odds of 125–1, on this happening, and pocketed £25,000 ($49,000) when it did in a match against Luton Town. "I've never placed such a large bet before" Mr. Hayward told the press afterwards, "but I had a feeling about it." 🗔

eld in May 2005, the Massachusetts Institute of Technology's Time Traveler Convention was, according to its organizers, a "mixed success." While a good time was had by all, no attendees admitted to being temporal vagrants; though this doesn't necessarily mean that there weren't any there.

One name that may have been on the guest list was John Titor. In November 2000, someone calling themselves "TimeTravel_0" appeared on the now defunct Time Travel Institute's web forum, commenting on a post about the "grandfather paradox" of time travel—that altering history may erase your own future existence. "The basics for time travel," he wrote, "start at CERN in about a year and end in 2034 with the first 'time machine'... Too bad we can't post pictures or I'd show it to you."

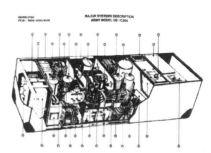

The forum's rules were swiftly amended and Titor did post his pictures, of a *Back to the Future*-style modified car interior and a handy time machine manual, and began holding forth on the ins and

outs of time travel. He claimed to be midway through a return trip from 2036 to 1975, where he had picked up an IBM 5100 computer that he planned to run antiquated software on. "My 'time' machine," he told another forum, "is a stationary mass, temporal displacement unit manufactured by General Electric. The unit is powered by two, top-spin, dual-positive singularities that produce a standard, off-set Tipler sinusoid."

Over the following months Titor held his own against skeptical physicists and people curious about the future, describing the subtle differences between ours and his own "worldlines"—i.e. parallel universes. "I've noticed little things like news events that happen at different times, football games won by other teams... things like that."

Titor provided dire warnings about our future: "There is a civil war in the United States that starts in 2005. In 2015, Russia launches a nuclear strike against the major cities in the United States, China and Europe. The United States counter attacks." Over 3 billion people are killed in this war, and by Titor's time, America exists as a system of small, self-sustaining feudal communities with extremely limited technological and energy reserves—hence the need for the lo-fi IBM 5100 computer.

On March 24, 2001 Titor vanished, as quietly as he had appeared. World affairs soon outpaced his predictions: the events of September 11, 2001 and the subsequent wars having been strangely absent from his prognostications.

While Titor's physics may have been up to speed, his history, unfortunately, wasn't. 🖵

100 RETROACTIVE PK: CAN WE CHANGE THE PAST?

The phenomenon of precognition—foreseeing events before they happen, while conscious or in dreams—has been recognized for centuries. On rare occasions, precognitive flashes have allowed people to change the future; but what if they could alter the past?

While astrophysicists dream of taming wormholes, parapsychologists have already produced what they consider to be tangible evidence that we can, at least on a very small level, reverse the process of cause and effect simply by thinking about it. Known as Retroactive Psychokinesis (RPK), it is probably the most puzzling of all lab-recorded psi-phenomena, yet it is also one of the most rigorously and comprehensively tested.

In the mid-'70s, Helmut Schmidt began to translate the output of Random Number Generators (RNG)—the equivalent of heads or tails in a coin toss—into audible clicks recorded onto audiotape. For the experiment, the RNG was run and its output recorded with nobody in the room to hear it. A copy was made of the audiotape and the original locked in a safe. Sections of the tape were then cut into test and control portions and, days later, the test portion of the copy given to an experimenter. This experimenter (who may or may not be aware that they are listening to pre-recorded clicks, it doesn't much affect the results) was instructed to listen to the tape and concentrate on increasing the frequency of clicks.

A statistically highly significant number of the results showed that, following the experiment, the test portions of the locked-up master tape contained considerably more clicks than the controls. The test subjects appeared to have altered RNG outputs made in the past!

The results of these experiments defy all common sense, yet have been successfully replicated several times by different researchers using a variety of RNG methods.

Esoteric branches of both quantum physics and information theory (where retrocausality is whispered of by physicists when they think nobody else is listening) have been invoked to attempt to make sense of these results. However, they still present an affront to rationality and so are invariably filed under "deep fringe" science. But if they are accurate—and many parapsychologists consider them to be amongst the most carefully controlled in the history of their field—the RPK experiments must throw into question our assumptions about causality, the flow of time and our experience of the reality around us. ⌑

101 DAMNED SCIENCE

"I can conceive of nothing, in religion, science or philosophy, that is anything more than the proper thing to wear, for a while," wrote the great American satirist of science and collator of "damned data," Charles Hoy Fort.

Born into a grocer's family in Albany, New York, in 1874, Fort learned quickly to distrust authority in all its forms, often getting into trouble for his mischievous practical jokes, such as swapping the labels on tins of fruit and vegetables—an early expression, perhaps, of his later rejection of scientific reductionism and compartmentalization: "one measures a circle beginning anywhere."

After youthful travels through the UK and South Africa, in 1897 Fort settled in New York's Bronx neighborhood to become a writer. Of ten novels only one, *The Outcast Manufacturers*, was published, in 1906. Fort and his wife Anna endured desperate poverty during this time, resorting to burning their furniture to keep warm. In moments of deep despair Charles also burned his unpublished manuscripts, reams of notes and thousands of newspaper clippings.

Fort spent much of his time scouring scientific journals and newspapers for what he called "damned data": records of unusual phenomena that fell outside, or contradicted, current scientific understanding. His first collection of this material, *The Book of the Damned* (1919) was rapturously received by intellectual and bohemian critics in the US, and led to the formation of a Fortean Society by one of his greatest supporters, the author and actor Tiffany Thayer. Fort, naturally, refused to become a member.

In *The Book of the Damned*, and three subsequent books, *New Lands* (1923), *Lo!* (1931) and *Wild Talents* (1932) Fort documents an incredible range of anomalous phenomena. Creatures fall from the sky, objects and people appear and disappear without warning

(Fort coined the term "teleportation"), spontaneous fires spark without warning, bizarre creatures stalk the Earth and strange lights fill the oceans and skies. It's a startling, baroque vision of the world, yet one drawn from the journals and newspapers that informed both science and society.

Fort marshaled his data into formidable "battalions of the accursed," leading a complex, sharply humorous assault on the bastions of science. He was equally dismissive of superstition and religious beliefs: "the fate of all explanation," he wrote, "is to close one door only to have another fly wide open."

Fort died in 1932, but his philosophy lives on, and we continue to live in a resolutely fortean universe: the more we learn about it, the stranger and more wonderful it becomes. ▢

Mark Pilkington is a freelance journalist, writer and editor whose work has appeared in the Guardian, the *London Times*, *Fortean Times*, *Arthur* and *The Wire*, amongst others. He also edits the highly-praised anthology of cultural marginalia, *Strange Attractor Journal* and runs Strange Attractor Press (www.strangeattractor.co.uk). He lives in London.